Regulation of Gastrointestinal Mucosal Growth

Integrated Systems Physiology: from Molecule to Function to Disease

Physiology is a scientific discipline devoted to understanding the functions of the body. It addresses function at multiple levels, including molecular, cellular, organ, and system. An appreciation of the processes that occur at each level is necessary to understand function in health and the dysfunction associated with disease. Homeostasis and integration are fundamental principles of physiology that account for the relative constancy of organ processes and bodily function even in the face of substantial environmental changes. This constancy results from integrative, cooperative interactions of chemical and electrical signaling processes within and between cells, organs, and systems. This eBook series on the broad field of physiology covers the major organ systems from an integrative perspective that addresses the molecular and cellular processes that contribute to homeostasis. Material on pathophysiology is also included throughout the eBooks. The state-of the-art treatises were produced by leading experts in the field of physiology. Each eBook includes stand-alone information and is intended to be of value to students, scientists, and clinicians in the biomedical sciences. Since physiological concepts are an ever-changing work-in-progress, each contributor will have the opportunity to make periodic updates of the covered material.

Published titles

(for future titles please see the Web site, www.morganclaypool.com/page/lifesci)

Regulation of Gastrointestinal Mucosal Growth
Jaladanki N. Rao and Jian-Ying Wang
www.morganclaypool.com

ISBN: 9781615041329 paperback

ISBN: 9781615041336 ebook

DOI: 10.4199/C00028ED1V01Y201103ISP015

A Publication in the Morgan & Claypool Publishers Life Sciences series

INTEGRATED SYSTEMS PHYSIOLOGY: FROM MOLECULE TO FUNCTION TO DISEASE

Lecture #15

Series Editors: D. Neil Granger, LSU Health Sciences Center, and Joey P. Granger, University of Mississippi Medical Center

Series ISSN

ISSN 2154-560X print

ISSN 2154-5626 electronic

Regulation of Gastrointestinal Mucosal Growth

Jaladanki N. Rao and Jian-Ying Wang
University of Maryland School of Medicine and Baltimore Veterans Affairs Medical Center

INTEGRATED SYSTEMS PHYSIOLOGY: FROM MOLECULE TO FUNCTION TO DISEASE #15

ABSTRACT

The mammalian gastrointestinal mucosa is a rapidly self-renewing tissue in the body, and its homeostasis is preserved through the strict regulation of epithelial cell proliferation, growth arrest, and apoptosis. The control of the growth of gastrointestinal mucosa is unique and, compared with most other tissue in the body, complex. Mucosal growth is regulated by the same hormones that alter metabolism in other tissues, but the gastrointestinal mucosa also responds to a host of events triggered by the ingestion and presence of food within the digestive tract. These gut hormones and peptides regulate the growth of the exocrine pancreas, gallbladder epithelium, and the mucosa of the oxyntic gland region of the stomach and the small and large intestines. Luminal factors (nutrients or other dietary factors, secretions, and microbes), which occur within the lumen and distribute over a proximal-to-distal gradient, are also crucial for the maintenance of the normal gut mucosal growth and could explain the villous height–crypt depth gradient and variety of adaptations since these factors are diluted, absorbed, and destroyed as they pass down the digestive tract. Recently, intestinal stem cells and polyamines are shown to play an important role in the regulation of gastrointestinal mucosal growth under physiological and various pathological conditions. In this chapter, we highlight key issues and factors that control gastrointestinal mucosal growth, with special emphasis on the mechanisms through which epithelial renewal is regulated by polyamines at the cellular and molecular levels.

KEYWORDS

epithelial renewal, proliferation, growth arrest, apoptosis, mucosal atrophy, polyamines, intestinal stem cells

Contents

Introduction

The epithelium of mammalian gastrointestinal (GI) mucosa has the most rapid turnover rate of any tissue in the body, and maintenance of its integrity depends on a complex interplay between processes involved in cell proliferation, differentiation, migration, and apoptosis. Under physiological conditions, undifferentiated epithelial cells continuously replicate in the proliferating zone within the crypts and differentiate as they migrate up the luminal surface of the colon and the villous tips in the small intestine. To maintain stable numbers of enterocytes, cell division must be dynamically counterbalanced by apoptosis, a fundamental biological process involving selective cell deletion to regulate tissue homeostasis. Apoptosis occurs in the crypt area, where it maintains the critical balance in cell number between newly divided and surviving cells and at the luminal surface of the colon and villous tips in the small intestine, where differentiated cells are lost. This rapid dynamic turnover rate of the intestinal epithelium is highly regulated at different levels and is critically controlled by numerous factors.

Consistent with most other tissues in the body, the regulation of the growth of GI mucosa is unique and affected by the same hormones such as insulin, growth hormone, thyroxine, and cortisol that alter metabolism in other tissues. However, the GI mucosa also responds to a host of events triggered by the ingestion and presence of food within the digestive tract. Food directly interacts with the GI mucosa and results in the release of several gut hormones that specifically affect only tissues of the GI tract and regulate its growth. Compared with gut hormones, various factors theorized to affect the mucosa within the lumen are less well defined. These factors include secretions (especially those of the pancreas and liver), luminal nutrition, and additional dietary constituents that stimulate growth independent of their nutritive value. Although physiological significance of these factors remains to be fully investigated, the common property of these stimulants, which accounts for their scientific interest, is that they occur within the lumen, distributed over a proximal-to-distal gradient. These luminal factors are diluted, absorbed, and destroyed as they pass down the GI tract, therefore, their effects on the mucosa also decrease distally. Such a mechanism has been thought to explain the villous-height–crypt-depth gradient and a variety of adaptations that occurs after exposure of portions of the gut to luminal contents, such as GI bypass surgery.

In this chapter, we will give an overview of the new advance in the regulation of GI mucosal growth, particularly the involvement of intestinal stem cells in epithelial renewal and cellular signals that are activated during adaptation and control epithelial cell division and apoptosis. We also examine the trophic properties of a variety of gut peptides and luminal factors and some of the hypotheses invoked to explain the gradient-oriented nature of mucosal growth. Finally, to summarize part of the knowledge in the area, we review the roles of cellular polyamines in GI mucosal growth and analyze in some detail the mechanisms by which polyamines regulate expression of various growth-related genes at both transcription and posttranscription levels.

• • • •

Intestinal Architecture and Development

The architecture of GI tract and its developmental features of different segments have been well defined. Several excellent review articles in this area are readily available [1–5]. Here, we only provide a brief overview on GI architecture and its developmental aspects that are relevant to our understanding of GI mucosal growth and regulation.

MUCOSAL WALL ARCHITECTURE

The primary functions of GI tract include digestion and absorption of nutrients, secretions, and immunoresponse. The unique architecture of the GI tract facilitates these functions, including multiple levels of infolding that result in an immense surface area, thus allowing maximal nutrient absorption. The wall of the intestine is conventionally described in terms of its component layers, and these layers are not separated entirely from one another but are joined together by connective tissue and by the neural and vascular elements. All segments of the GI tract are divided into four layers: the mucosa (epithelium, lamina propria, and muscular mucosae), the submucosa, the muscularis propria (inner circular muscle layer, intermuscular space, and outer longitudinal muscle layer), and the serosa (Figure 1). Mucosa is the innermost layer, which is structurally and functionally the most complex and important area. The mucosal surfaces of the body are the areas where important absorptive function occurs. The mucosa consists of three layers. The first layer facing the intestinal lumen is made up of epithelial cells, which is a single layer in the GI tract and is attached to a basement membrane overlying the second layer, the lamina propria, which consists of subepithelial connective tissue and lymph nodes, underneath which is the third and deepest layer called muscularis mucosae. This is a continuous sheet of smooth muscle cells that lies at the base of the lamina propria. The entire mucosa rests on the submucosa, beneath which is the muscularis propria. The outermost layer is named as the serosa or, if it lacks an outer layer of mesothelial cells, the adventitia. The submucosa consists of a variety of inflammatory cells, lymphatics, autonomic nerve fibers, and ganglion cells. This area is also a branching and distribution zone for arteries and small venous channels.

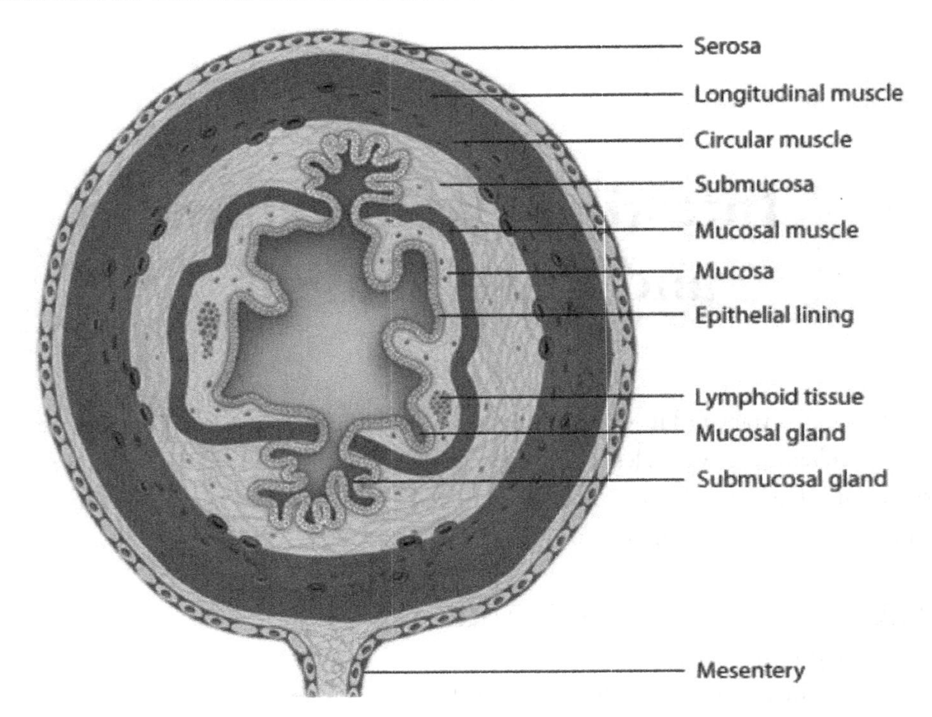

FIGURE 1: Architecture of the gut mucosal wall. Four-layered (mucosa, submucosa, muscularis mucosa, and serosa) organization of the digestive tract. Adapted from http://www.virtualmedicalcentre.com (permission pending).

In the GI tract, the muscularis propria contains smooth muscle cells organized into a tightly coiled, inner circular layer and outer longitudinal layer, as shown in Figure 1. The smooth muscle cells are arranged in parallel arrays. Between the outer and inner layers of the muscularis propria are prominent autonomic neural fibers and ganglionic clusters that form a myenteric plexus. The major functions of the muscularis propria are to propel food through the gut by contractile peristaltic waves initiated and regulated by various neural and hormonal events [2,6]. Flow is regulated by peristaltic mechanisms and by sphincters located in the upper esophagus, in the distal portions of the esophagus, stomach, and ileum and in the anus. Most part of the intestine is lined on its outer surface by a sheath of protective layer, the serosa, which consists of a continuous sheet of squamous epithelial cells, the mesothelium, separated from the underlying longitudinal muscle layer by a thin layer of loose connective tissue (Figure 1). The serosal layer forms a natural barrier from the spread of inflammatory and malignant processes [7,8].

DEVELOPMENT AND FUNCTIONS

The development of the mammalian GI system is preprogrammed, but this can be altered during the intrauterine and early postnatal life [6,9,10]. There are two major steps involved in the development of the GI tract, formation of the gut tube and formation of the individual organs with their specialized cell types. Genes regulating both phases are being identified and well characterized in published comprehensive reviews [6,11,12].

Esophagus

The esophagus is the foremost part of the GI tract that can be identified as a distinct structure early in the human embryogenesis. This organ elongates during subsequent development relatively more rapidly than the fetus as a whole [2,6,13]. The major events during the formation of esophagus are as follows: at 10 weeks, ciliated columnar epithelium appears followed by the replacement of stratified squamous epithelium at around 20–25 weeks, a process that begins in the mid-esophagus and proceeds further [14]. Studies by Hitchcock et al. [9] show the development of esophageal musculature and innervation in fetuses at 8–20 weeks of gestation and in infants at 22–161 weeks of age. The esophagus is well supplied with lymphatics that form a richly anastomosing network in the lamina propria and submucosa. Although the esophagus is described as a tube, it is oval and has a flat axis anterior to posterior with a wider transverse axis. The primary functions of the normal esophagus are the propulsion of food from the mouth to the stomach and the prevention of significant reflux of gastric contents into the esophagus. The propulsive function is affected by involuntary peristalsis in the muscularis propria that unlike the remainder of the GI tract is formed by two types of muscle fibers, such as striated and smooth muscles [2,15]. When it is on the resting state, the esophagus is a collapsed tube, and the elastic tissue in its walls accounts for its distensability. During swallowing, the lumen dilates, and the folds flatten so that the esophagus can normally accommodate the passage of even large amounts of food bolus.

Stomach

The stomach receives food from esophagus and is a J-shaped reservoir of the digestive tract, in which ingested food is soaked in gastric juice that contains digestive enzymes acids [2,12]. The prenatal ultrasound examinations have revealed that the stomach grows in a linear fashion from 13 to 39 weeks and that the characteristic anatomic features, such as greater curvature, fundus, body, and pylorus, are identified in as early as 14 weeks [6]. The stomach is located in the left upper quadrant of the abdomen, and its upper portion lies beneath the dome of the left hemi-diaphragm. The stomach is divided into four zones, each of which has a specific microscopic mucosal structure. The "cardia" is the narrow portion of the stomach immediately distal to the gastroesophageal junction. The remainder of the stomach is divided into proximal and distal parts. The proximal portion is the body or corpus, and the distal part is named as pyloric antrum which is demarcated from the duodenum by the pyloric sphincter. The pyloric sphincter is closed in the resting state to prevent the reflux of intestinal contents into the stomach (Figure 2). The arterial blood supply to the stomach involves many different branches, among which, splenic artery, common hepatic, and left gastric arteries are important. Venous drainage from the stomach is through the portal system to the liver. Deeper in

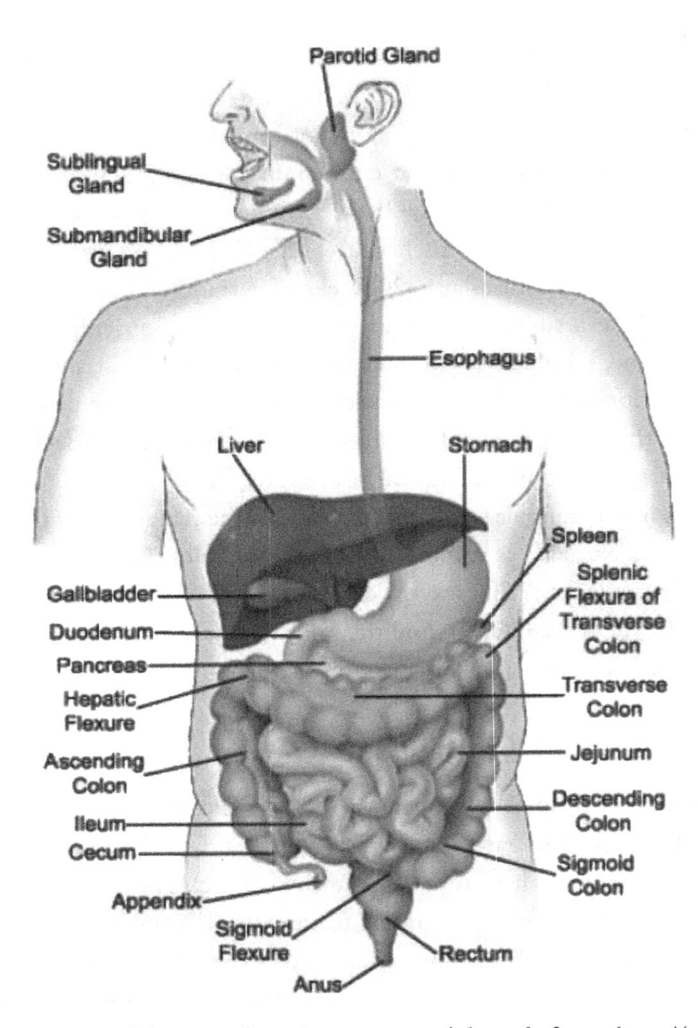

FIGURE 2: Architecture of human digestive system. Adapted from http://www.vitallywell.net/digestive-enzymes.html (permission pending).

the epithelial wall is a rich lymphatic network that drains to the regional perigastric lymph nodes and to the nodes in the omentum, around the head of the pancreas and in the spleen [16]. In the stomach, solid food is fragmented and mixed by peristalsis. A semiliquid material (chime) is formed and released in small, regulated bursts into the duodenum by rhythmic openings of the pyloric sphincter. Cells in the corpus and fundus of the stomach also produce hydrochloric acid and intrinsic factors necessary for absorption. Although it occurs predominantly in the small intestine, some digestion occurs in the stomach. Certain gastric mucosal cells produce pepsinogens, the proteolytic enzymes that are secreted in an inactive form, but they are then activated by the acid environment of the gastric lumen during food intake. In addition, production of the hormone gastrin is also another major gastric function. The development of gastric glands (fundic type or oxyntic) occurs very early during human fetal life (10–12 weeks of gestation) [6,17,18]. The advance and detailed description of development, gastric endocrine cells, and functions of the stomach are described in recent review articles [6,13,19].

Small Intestine

The intestinal tract followed by the stomach consists of the small intestine, including duodenum, jejunum, and ileum, and the large intestine or colon (Figure 2). The development of small intestine consists of three successive phases: morphogenesis and cell proliferation, cell differentiation, and cellular and functional maturation [6,20]. Organogenesis of the intestine is completed by 13 weeks of gestational period [21]. The duodenum is the first portion of the small intestine and extends approximately 25 to 30 cm from the pyloric sphincter to a fibrous and muscular band, the ligament of Treitz. From the distal part of the stomach, duodenum enters the retroperitoneum, curves, and then enters into peritoneal cavity. Jejunum and ileum are located between the Trietz and the ileocecal sphincter. First one-third of this segment of the small intestine is referred as the jejunum, whereas the remainder is named as the ileum. Structure and function of jejunum and ileum are different and occur gradually during the development. The blood supply for the three segments of small intestine derives from the celiac, superior, and inferior mesenteric arteries, respectively. The cecal and appendiceal diverticulum appear during 6 weeks of gestation, marking the margin between the small and large intestines. The inner surface of the small intestine is covered with a simple columnar epithelium exhibiting invaginations, known as the crypts of Lieberkuhn, which are comprised predominantly of proliferating cells, and finger-like projections called villi that contain the majority of differentiated absorptive cells [2,6,19]. The epithelial lining initiates and modulates the basic activities attributed to the small intestine, like terminal digestion of nutrients and transport of nutrients, water, and ions. The epithelial surface is expanded by villous thickness and crypts present between villi. The adult small intestinal epithelium is composed of four different cell lineages. Differentiated cells, such as enterocytes, enteroendocrine, and goblet cells, occupy the villi, while another type of differentiated cells, the Paneth cells, reside at the bottom of the crypts and secrete antimicrobial agents. The remaining part of the crypts constitutes the stem cells and proliferating progenitor compartment [20,21].

Large Intestine

The large intestine or the colon arches around the small intestine, commencing in the right ileac region. In adult humans, the colon is approximately 1.5 m in length. The parts of the large intestinal anatomic divisions from proximal to distal end include the cecum, ascending colon, hepatic flexure, transverse colon, splenic flexure, descending colon, sigmoid colon, rectum, and anus. The structure of the colon in many respects overlaps that of the small intestine as described previously [2,6,22]. Development of the colon is marked by three important cytodifferentiative stages, which include the appearance of primitive stratified epithelium to a villous architecture with developing crypts at about 12–14 weeks of gestation and followed by the remodeling of the epithelium at around 30 weeks when the villi disappear and the adult-type crypt epithelium is established [6]. Concurrent with the presence of villous morphology, the colonic epithelial cells express differentiation markers similar to those in small intestinal enterocytes [22]. As seen in the small intestine, lymphoid nodules that distort the normal mucosal architecture are present in the colon, and the colonic epithelium also rapidly renews by itself. Undifferentiated crypt cells appear to be the progenitor for all cell types in the colon. In contrast to the small intestine, the mucosa of the large intestine is not covered with villous projections but it contains deep tubular pits that increase in depth toward the rectum and extends as far as the muscularis mucosa. Colonic mucosal epithelial cells include absorptive cells, goblet mucus cells, undifferentiated columnar crypt cells, caveolated cells, Paneth cells, and M-cells present in the colonic mucosa and are almost identical to those cells present in the small intestine. The major functions of the colon are to reclaim luminal water and electrolytes.

• • • •

Characteristics of Gut Mucosal Growth

The gastrointestinal epithelium is a complex system that consists of multiple cell types undergoing continual renewal while maintaining precise interrelationships [23–28]. Under biological conditions, a huge number of cells are exfoliated regularly throughout the lumen of the GI tract, and these sloughed cells are almost immediately replaced by new cells from the stem cell compartment. These stem cells or proliferative progenitors in the crypts generate epithelial cells that differentiate during their migration toward the villous region. A wide variety of dietary and growth factors and hormonal and transcriptional factors are involved in the regulation of GI mucosal growth and development [24,25,28]. A well-controlled cascade of signals maintains mucosal growth by the shedding of senescent and apoptotic cells at the surface of the GI epithelium [26,28,29]. The overall mucosal integrity depends on the dynamic balance between cell production and cell loss. A defect in cell production leads to the development of mucosal atrophy and results in a decrease in absorption, whereas mucosal hyperplasia results from excess production of newborn cells. Hyperplasia can cause hypersecretion and increase the risk of cancers [10,27,29].

The mucous neck cells are located throughout the stomach, predominantly in the upper portion (neck or isthmus) of each gland, immediately below the glandular junction [2,27] and are considered as gastric epithelial progenitor cells. In each gland, the mucous neck cells form the zone of epithelial cell renewal, giving rise to new surface faveolar mucous cells as well as the other cell types within the glands. It has been thought by many investigators for years that parietal (oxyntic) cells are unable to divide [30] and are replaced by newly formed cells migrating slowly down the gland and differentiating into acid-producing cells [31]. Based on the results obtained from cultured human gastric corpuses from 12 to 17 weeks of gestation, the proliferative compartment of the mucosa from which the differentiated cells arise is the neck portion at the base of the glandular compartments [2,6]. Most parietal cells occupy the mid portion of the gastric glands and are important in the processes of differentiation and development of the stomach. In the mouse, parietal cells survive for ~90 days, which is the time period for migration to the bottom of the gland [27,32]. Another type of cells located in the gastric glands is zymogen (chief) cells which are concentrated at the base of the gastric glands. After injury, zymogen cells are derived from the stem cells, but they are replaced by the process of mitosis under normal circumstances [27,30]. In addition, several specialized endocrine-secreting cells are scattered in the antral region of the stomach. G-cells in this region produce gastrin. The levels of antral gastrin are varied during development, nutritional intake, and various disease states [33]. Regulation of G-cell populations is important because they participate in the GI mucosal growth through the release of gastrin that induces the growth of the oxyntic mucosa. In this regard, transgenic mice overexpressing gastrin exhibit an increased cell proliferation. In contrast, mice with a targeted deletion of gastrin display a decreased number of mature parietal cells with an increase in the number of mucous neck cells [34]. [^{3}H]-thymidine incorporation studies have revealed that G-cells are derived by the mitosis cell division from other G cells [35], whereas gastrin infusion stimulates cell proliferation in the neck region of the oxyntic gland where gastric stem cells may reside. It is likely that gastrin interacts with gastric stem cells to stimulate their proliferation and differentiation into parietal cells.

The proliferative compartment of the small intestinal epithelium is structured as several bottle-shaped invaginations known as crypts of Lieberkuhn. Each of the intestinal crypts is a developmental unit that contributes to the renewal of the intestinal epithelium when growth progresses to adulthood. The crypts contain the crypt base columnar cells (CBC), believed to be intestinal stem cells (as discussed separately in another chapter), and slowly duplicate and eventually produce transient population of progenitor cells. The majority of epithelial cells migrates upward to the top of the villi and differentiates into endocrine, goblet cells, and enterocytes. Enterocytes responsible for the secretory and absorptive functions of the gut epithelium are the most abundant. The population of mucous producing goblet cells and endocrine cells makes up ~5% and ~1% of the total epithelial cells [2,6]. This process occurs continuously, cells are replaced at the tip of the villous every 4 to 5 days. This phenomenon in neonatal piglets, which enhances mitosis, is paralleled with a decrease in apoptosis at the first 2 days after birth, resulting in transient elevation of the mitosis/apoptosis ratio, which contributes to the enlargement of the gut mucosa [19]. GI mucosal cell growth is regulated by a number of nutritional and hormonal factors. For example, deprivation of food results in decreased cell proliferation, which is reversed by refeeding [13,26,28]. An interesting study has shown that ectopic expression of Cdx2 gene induces intestinal epithelial cell differentiation [36,37]. Several signaling pathways are shown to play an important role in patterning the gut during development and in regulating epithelial differentiation. Large intestine has no Paneth cells and villi; therefore, it has a relatively flat surface. The upward migration of cells from crypts ends with their incorporation into surface epithelium cuff, and each cuff is supplied by one crypt, unlike in the small intestine, where each villous receives cellular outputs from several surrounding crypts [6,38]. Differentiated cells in the colon are also derived from CBC, and their turnover time in colonic mucosa is ~3 to 4 days.

• • • •

Intestinal Stem Cells

The continuous renewal and repair of adult intestinal mucosal epithelium after injury depend on resident specialized stem cells. Stem cells are cells that are capable of self-renewal and consistently maintain themselves over long periods of time, producing all undifferentiated cell types of that tissue. In various tissue types, stem cells serve as a sort of internal repair system, dividing essentially without limit to replenish other cells as long as the organism is alive [39–42]. An increasing body of evidence shows that each adult tissue contains its own unique type(s) of dedicated stem cells; there are many different types of stem cells in mammals. To date, it remains largely unknown if shared molecular and cell biology principles underlie the behavior of various tissue-specific stem cells. Generally, stem cells were thought to come from two main sources, embryos formed during the blastocyst phase of embryological development (embryonic stem cells) and nonembryonic "somatic" or "adult." Both types of stem cells are typically characterized by their potency or potential to differentiate into another type of cells with more specialized functions, such as muscle cells, red blood cells, brain cells, or epithelial cells. Stem cells are distinguished from other cell types by two important characteristic features: (1) stem cells are unspecialized cells capable of renewing themselves through cell division, sometimes after long periods of inactive state and (2) stem cells can be induced to become tissue- or organ-specific cells with specialized biological functions under certain physiological or experimental conditions. In several organs, such as gut mucosa and bone marrow, stem cells regularly divide to repair and replace worn out or damaged cells, while in some other organs, including pancreas, muscle, brain and the heart, discrete populations of adult stem cells generate for replacements of cells that are lost through normal wear and tear, injury, or disease. Given their unique regenerative capabilities, stem cells are able to offer novel and new potential therapeutic approaches for treating different diseases, such as diabetes and heart disease.

The intestinal epithelial villus/crypt structure, its surrounding pericryptal fibroblasts, and mesenchyme form an anatomical unit that generates four cell lineages, namely absorptive enterocytes and goblet, enteroendocrine, and Paneth cells of the secretory lineage. The crypt is a contiguous pocket of epithelial cells at the base of the villus, in which intestinal stem cells (ISCs) are periodically activated to produce progenitor or transit amplifying (TA) cells which are committed to produce mature cell lineages [43,44,45]. In normal conditions, newly formed TA cells reside within the crypts for 2–3 days and undergo up to six rounds of cell division. When these newly divided cells reach the crypt–villus junction, they rapidly differentiate into each of the four terminally differentiated cell types (Figure 3). The crypt is mainly occupied by undifferentiated cells; but differentiated Paneth cells that secrete antibacterial peptides into the crypt are also unusually located at the base

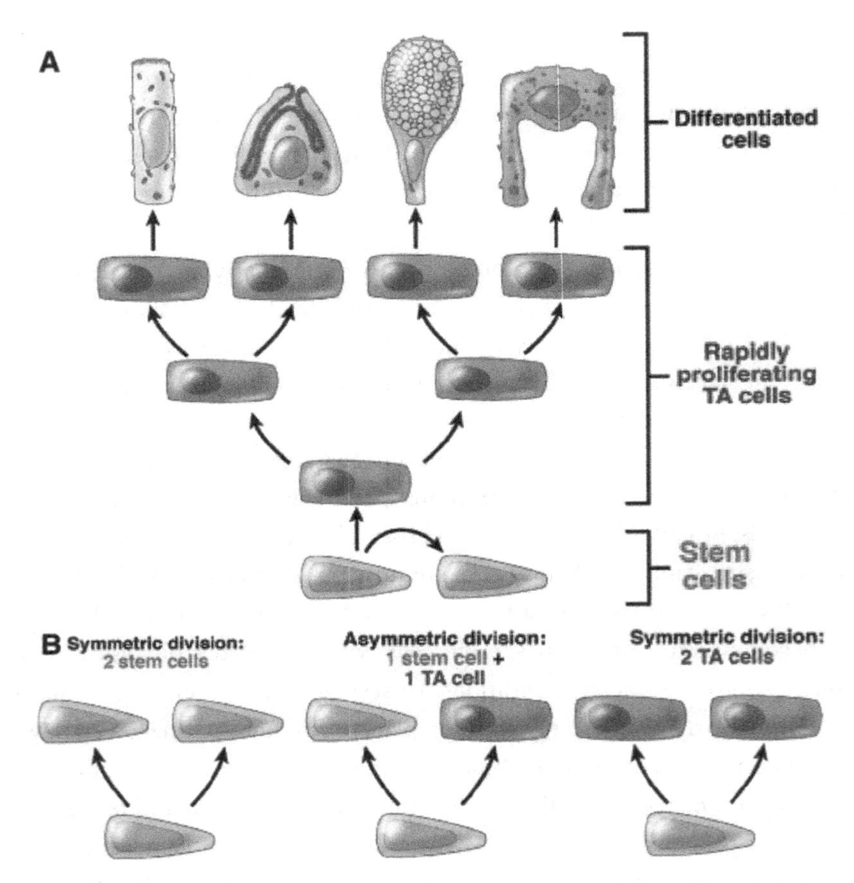

FIGURE 3: A scheme of adult stem cell-driven tissue renewal in organs such as intestine and stomach. (A) Stem cells concomitantly self-renew and generate, rapidly dividing transit-amplifying (TA) daughter cells via asymmetric cell division. The TA cells undergo several rounds of division before differentiating into the mature, functional cell types of the adult tissue. (B) Adult stem cells potentially can follow 3 modes of cell division: symmetric division to generate 2 stem cells, asymmetric division to generate 1 stem cell and 1 TA cell, or symmetric division to generate 2 TA cells. Used with permission from *Gastroenterology*: 138: pp. 1681–1696, 2010.

of the crypt area and escape their upward migration [45,46]. It is likely that ISCs divide asymmetrically, thus giving rise to one daughter stem cell and one TA cell, which differentiate toward mature epithelial cells. In the damaged gut mucosa, after exposure to irradiation or chemotherapy, ISCs are undergoing symmetrical division that yield into two stem cells to replace damaged ISCs. It has been shown that there are several intrinsic and extrinsic cellular mechanisms that regulate ISC self-renewal and differentiation of newly divided cells.

ISCs AND THEIR NICHES

The ISC niche is an anatomical structure that is composed of stem cells, their progeny, and elements of their microenvironment, and it coordinates the normal homeostatic production of functional mature cells [38,47,48]. ISC niches provide a sheltering environment that sequesters ISCs from various stimuli, such as differentiation and apoptosis and are also safeguarded against excessive ISC production, thus reducing the risk of cancers. The epithelial homeostasis of the intestine is based on a delicate balance between self-renewal and differentiation. Intestinal pericryptal fibroblasts (also named as subepithelial myofibroblasts) are also implicated in the formation of ISC niches, and they are believed to secrete various putative growth factors and cytokines that promote epithelial proliferation and enhance production of differentiated cell lineages, including enterocytes, goblet, enteroendocrine, and Paneth cells (Figure 4). There are two models to explain how pluripotent stem cells fuel the proliferative activity of crypts [45,49,50]. One concept is the "+4 position" model from the crypt bottom and the other is called the "stem cell zone" located below the +4 position (Figure 4). Each of the crypts is commonly believed to contain approximately four to six independent stem cells. BrdU-labeling studies suggest that label-retaining cells reside specifically at the +4 position relative to the crypt bottom, with the first three positions occupied by the terminally differentiated Paneth cells and that +4 cells are extremely sensitive to radiation, a property that functionally protects the stem cell compartment from genetic damage. More support for the +4 model also results from recent studies on lineage tracing experiments utilizing a newly generated Bmi-Cre-ER knock-in allele [51,52]. After induction for 24 h, the cells expressing Cre-reporter are located at the +4 position, directly above the Paneth cells. The model of "stem cell zone" was originally proposed by Cheng and Leblond [53], based on the identification of crypt base columnar (CBC) cells (Figure 4). These CBC cells are small undifferentiated cycling cells hidden between the Paneth cells [45]. Recently, Barker et al. further identified the Wnt target genes, Prominin and Lgr5/GPR49, as markers that specifically label CBC stem cells in the mouse small intestine [54,55]. It has been shown that CBC cells are capable of long-term maintenance of intestinal epithelial self-renewal and generate differentiated mature intestinal epithelial cells (Figure 4). Furthermore, lineage-tracing experiments show that both CBC and +4 cells produce offspring within days and persist for up to a year and that they are multipotent stem cells and exhibit different cycling kinetics and molecular features [45,56].

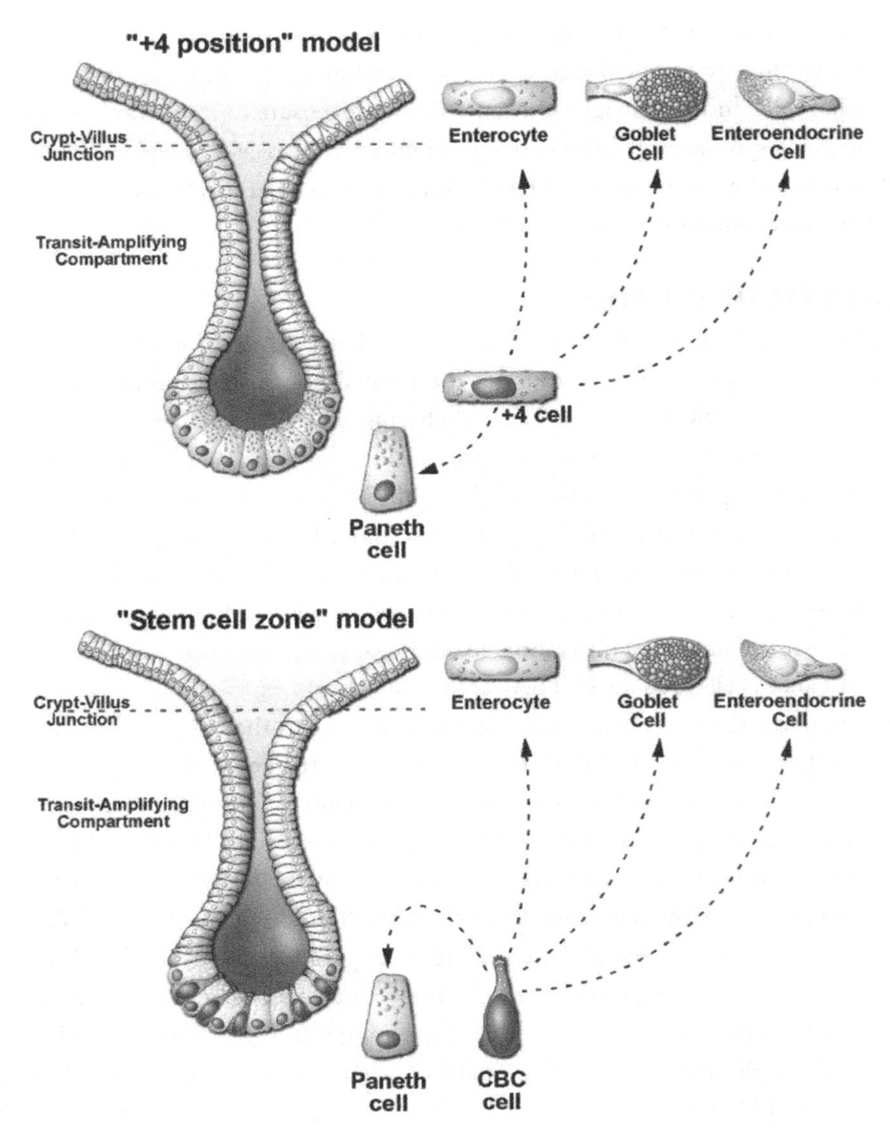

FIGURE 4: Two opposing models showing the identity of the intestinal stem cells. The exact identity of the intestinal stem cells has proven controversial over the last 30 years, with two opposing models dominating the literature. Top: "+4 position" model. It was assumed that the crypt base is exclusively populated by terminally differentiated Paneth cells, and the stem cells must therefore be located just above the Paneth cells at the +4 position. This model, largely championed by Chris Potten and colleagues,

It is possible that these two types of stem cells coordinately regulate intestinal epithelial tissue homeostasis and regeneration under physiological and various pathological conditions.

SIGNALING PATHWAYS REGULATING ISCs

Several studies using transgenic and knock-out animal models suggest that different signaling pathways, such as Wnt, bone morphogenic protein (BMP), Notch, Ephrin, JAK/STAT1, PTEN, AKT, and PI3K, play an important role in the regulation of intestinal epithelial renewal, particularly ISC function [43,46,51,57,58]. Recently, microRNAs (miRNA) are also shown to modulate ISC cell proliferation and differentiation [59,60]. Disruption of these pathways alters gut mucosal growth and could lead to intestinal mucosal tumorigenesis.

Wnt Signaling Pathway

The Wnt signaling pathway has a unique and central role in the regulation of intestinal epithelial renewal under biological conditions. It is not only the dominant force behind the proliferative activity in the crypts but also the principal cause of colon cancer after its mutational activation. The Wnt signaling is the first pathway that is shown to regulate ISC functions. Activation of the Wnt pathway is crucial for the maintenance of stem/progenitor cell division and for newly divided cell migration along the crypt–villus axis [61–63]. Several lines of *in vivo* evidence show that TA cell proliferation in the crypt is strictly dependent on the continuous stimulation of canonical Wnt signaling pathway, thus activating nuclear β-catenin/T cell factor (TCF) transcriptional activity [61]. In the absence of Wnt signals, free cytosolic β-catenin is sequestered and targeted for degradation via the β-catenin destruction complex (Figure 5).

It has been shown that crypt epithelial cells highly express Wnt receptors and co-receptors, such as frizzled and LRPs, and are targets of Wnt signals. Genetic experiments suggest that Wnt signals pattern the physical structure of ISC niches by generating opposing and complementary gradients of ephrins and their tyrosine kinase receptors [61,63,64]. Silencing TCF4 and β-catenin or inhibition of Wnt activity by overexpression of its natural inhibitor Dickkopf homologue-1 (Dkk-1) inhibits ISC proliferation in the intestinal epithelium, whereas mutations of APC (a negative regulator

---

predicts that the enterocytes, goblet cells, and enteroendocrine cells are derived from +4 cell progeny that differentiate as they migrate out of the crypts onto the villi. In contrast, the Paneth cells differentiate as they migrate down from the +4 position toward the crypt base. Bottom: "stem cell zone" model. Proposed by Leblond and colleagues states that small, undifferentiated, cycling cells (termed crypt base columnar cells) intermingled with the Paneth cells are likely to be the true intestinal stem cells. Definitive proof for either model has proven elusive due to the lack of specific markers for these cells. Used with permission from *Genes Dev* 22: pp. 1856–1864, 2008.

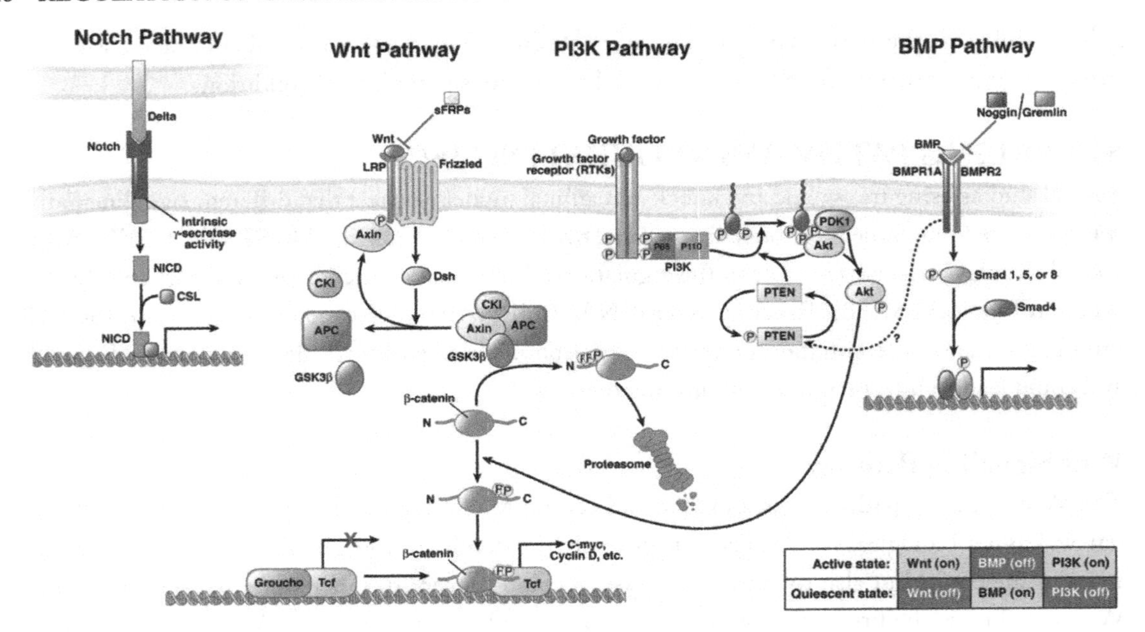

FIGURE 5: Signaling pathways within the crypt-ISC activation. Outline of Notch, Wnt, PI3K, and BMP pathways and their potential points of interaction. Normally, +4 LRCs are maintained in a quiescent state through canonical BMP signaling via the transcriptional effects of Smads and/or possible regulation of PTEN and subsequent inhibition of PI3K signaling. In addition, Wnt inhibitors, such as secreted frizzled-related proteins (sFRPs), act to hinder Wnt mediated effects. Transient activation of +4 LRCs is coordinately regulated by Wnt pathway activation, expression of BMP antagonists, such as Noggin, that abrogate BMP inhibition, as well as PI3K activity. Induction of the PI3K pathway results in Akt activation and subsequent C-terminal β-catenin phosphorylation which likely affects the nuclear activity of this molecule. Thus, Akt assists Wnt-induced β-catenin activation that promotes cell cycle entry and progression within these putative ISCs. Evidence also suggests that Notch pathway activation may be permissive for Wnt induced crypt cell proliferation. Used with permission from *Gastroenterology* 134: pp. 849–64, 2008.

of Wnt) or expression of oncogenic forms of β-catenin cause hyperproliferation (Figure 5). Besides mitogenic activity, Wnt signaling pathways are also involved in the regulation of differentiation of Paneth cells at the crypt bottom [65]. After the cells are generated near or at the crypt base, they migrate toward the villus while undergoing the process of maturation and differentiation. Ephrin molecules that play a role in the maintenance of cellular boundaries and migratory paths are identified as target genes of Wnt and are shown to segregate cells along the crypt–villus axis [64,65]. Wnt activation enhances expression of Ephrin B receptors and ligands through increasing β-catenin/TCF transcription complex within the intestinal epithelium.

Several recent studies conducted by Clevers and colleagues identify the specific markers of ISCs [38,51,65]. Although a great majority of the genes are expressed throughout the proliferative crypt compartment, the Lgr5/Gpr49 gene is expressed in a particularly unique fashion. The Lgr5 gene encodes an orphan G protein-coupled receptor (GPCR), characterized by a large leucine-rich extracellular domain; it is closely related to GPCRs with glycoprotein ligands, such as TSH, FSH, and LH receptors. Lgr5 is highly expressed in the stem cells of another Wnt-driven self-renewing structure, the hair follicle. APC^{min} mouse exhibit the expression of Lgr5 in a limited number of crypt bottom cells as well as in adenomas. It has been reported that single $Lgr5^{+}$ ISC regenerates the self-renewing and functional crypt-like structures *in vitro* when it is exposed to appropriate signaling factors and extracellular matrix [51]. Similar results are also observed by using Bmi, Prominin-1 and both Prominin-1 and Lgr5 [43,52,54,67]. Other lineage-tracing experiments also reveal that $Lgr5^{+}$ cells represent actively dividing and multipotent ISCs that contribute to the long-term renewal of the entire gut epithelium. The Lgr5 marker now allows identification of stem cells not only in the intestine but also in other tissues, such as hair follicle, mammary gland, and stomach epithelium. All results obtained from different tissues support the notion that $Lgr5^{+}$ cells represent a more general marker of adult stem cells [43,51,54,68]. In the mammalian intestinal stem cell system, several markers, such as Lgr5, Prominin 1, and Bmi, are currently available to trace stem cell lineage and stem cell identities. It has been noticed that colon cancer stem cells can be identified through these specific markers, including Lgr5, CD133 glycoprotein, Musashi-1, CD29, and CD24 and Lgr5 and can be used for targets for novel therapies in the future [51,69,70].

BMP Pathway

BMP is a member of transforming growth factor-β family and specifically binds to BMP receptors. BMP signaling plays an important role in the regulation of intestinal development and adult epithelial tissue homeostasis. BMP2 and BMP4 isoforms are expressed in the mesenchyme, and their receptors are identified in the epithelium [46,71,72]. Generally, BMP functions as a negative regulator of intestinal epithelial cell proliferation in the crypts; and conditional deletion of BMP receptor 1A results in cell hyperproliferation. The BMP antagonist "noggin" is also expressed in the submucosal region adjacent to the crypt base as well as dynamically in +4 cells, whereas overexpression of noggin causes ectopic crypt formation [46,73]. In addition, BMP-2 and BMP-4 null mice are embryonically lethal. Studies using conditional ablation of BMP receptor 1A suggest that BMP signals serve to antagonize the crypt formation and ISC self-renewal (Figure 5) [73,74]. Inactivation of BMP receptor-1A or overexpression of Noggin increases β-catenin nuclear translocation, suggesting that there is a cross-talk between BMP and Wnt pathways.

Notch Pathway

Like Wnt and BMP signals, the Notch pathway is also essential for maintaining the crypt compartment at undifferentiated and proliferative states during gut development [43,75–77]. Notch genes encode single transmembrane receptors that regulate a broad spectrum of cell fate decisions. Components of Notch pathway are expressed in the crypt base. Functions of Notch signaling in the regulation of ISC activity are revealed by studies using knock-out animal models. Activation of Notch signaling in the intestinal epithelium increases cell proliferation, but inhibition of Notch reduces secretory cells (Figure 5). Although Notch signaling plays a role in ISC proliferation, current literature suggests that it functions in the TA compartment controlling absorptive (enterocytes) rather than secretory (enteroendocrine, goblet, and Paneth cells) fate decisions in the intestinal epithelium [43,66].

Little is known about the involvement of other signaling pathways including PTEN, PI3K, JAK/STAT in the regulation of ISC proliferation, although direct or indirect evidence exists in the literature (Figure 5). Due to space limitations, this information is not mentioned here, but can be found in several recent excellent review articles [46,48,58,78–80].

• • • •

Role of GI Hormones on the Gut Mucosal Growth

GI hormones are chemical messengers that are implicated in many aspects of physiological functions of the gastrointestinal tract, including the regulation of secretion, absorption and digestion, and gut motility. GI hormones are a large family of peptides and are secreted by endocrine cells that are widely distributed throughout the GI mucosa and pancreas. Gastrin, secretin, and cholecystokinin (CCK) were the first discovered gut hormones, and as of today, there are more than 50 gut hormone genes and a multitude of bioactive peptides, which makes the gut as the largest endocrine organ of the body. GI hormones act through their specific receptors to activate particular signal pathways and ultimately provide functional signals for their physiological effects [81–84]. Initially, these hormones were described solely as endocrine products, but subsequent experiments have revealed that GI hormones also function as autocrine or paracrine to regulate GI functions. In addition, these hormones are thought to serve as transmitting agents for nerve impulses discharged into blood vessels after nervous stimulation in a true neurocrine fashion [85,86]. Besides their regulatory effects on secretion, absorption and digestion, and gut motility, GI hormones also modulate GI mucosal growth and are involved in the pathogenesis of gut mucosal atrophy, neoplasm, and cancers. The GI hormones that regulate gut mucosal growth positively or negatively include gastrin, CCK, secretin, somatostatin, ghrelin, bombesin, and gastrin-releasing peptide (GRP).

GASTRIN

Gastrin was first identified by Edkins in 1906 [87] when he discovered that extracts of antral mucosa stimulated acid secretion (gastric juice) from the gastric fundus, and then he named this new active agent within the antral extract as "gastrin." Almost 55 years later, Gregory and Tracy [88] successfully isolated the pure form of gastrin from the antral mucosa of hogs. Since then, gastrin has been established as the major biological regulator of gut physiology and plays a critical role in the regulation of gastric acid secretion [89–91]. Gastrin is initially released from the G cells in the antral region of the stomach during a meal by vagal stimulation, distention and digested protein. Other organs and cells that also produce gastrin include pancreatic endocrine cells [92], pituitary [93], and extraantral G cells [94]. The cellular targets for gastrin in the stomach are the acid-secreting parietal cells and histamine producing enterochromaffin-like (ECL) cells. In addition to stimulating acid secretion from gastric parietal cells, gastrin is also considered to be a key growth regulator in the gut mucosa and is implicated in the development of various GI cancers [91,95–97].

The major biologically active forms of gastrin are G-17 and G-34 amino acid peptides containing tyrosine residues at carboxyl terminus [96]. The biosynthetic pathways, leading to the production of amidated gastrins from the precursor molecule, prograstrin, are well established [82,84,98]. In antral G cells progastrin is stored and processed into secretory granules; N- and C-terminal extensions are removed by prohormone convertases. The posttranslational modification of gastrin appears to have no functional significance, as all forms are equally potent at the receptor level. G-17 is cleared from the circulation faster than the G-34 form, therefore, the majority of gastrin in the circulation during fasting is G-34. In contrast, the major form of gastrin that is released after a meal is the G-17 [99] (Figure 6). The greatest proportion of gastrin in the circulation is fully processed. As mentioned above, gastrin secretion from antral G cells is tightly regulated by luminal, paracrine, endocrine, and different neuronal stimuli [99,100]. Small peptides, aromatic amino acids, and calcium in a meal are also key contributors for the stimulation of gastrin release (Figure 6).

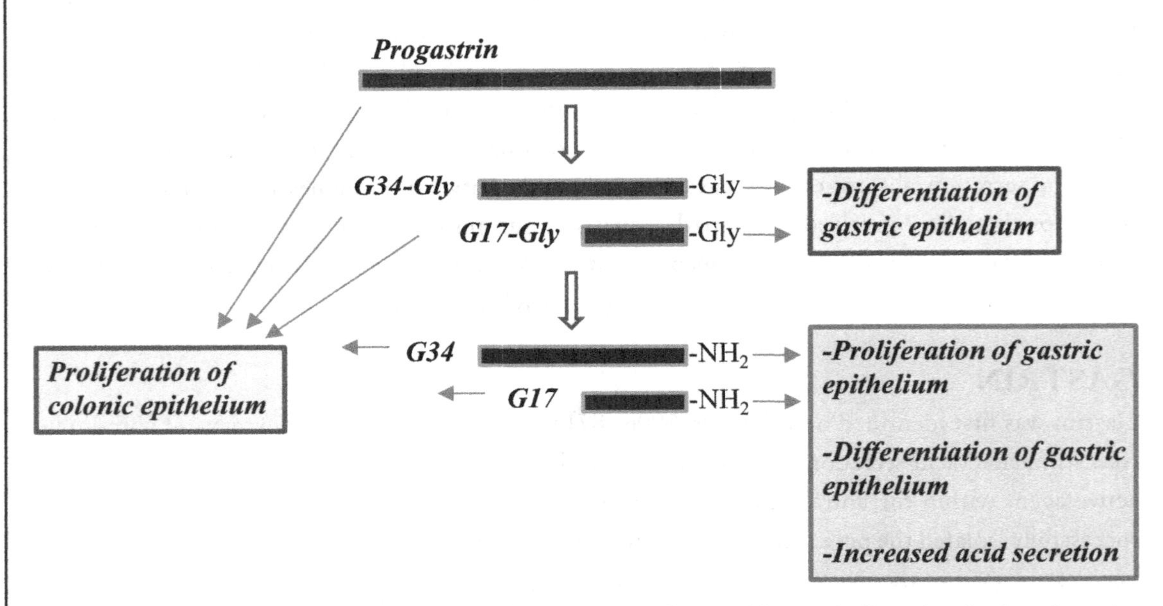

FIGURE 6: Role of gastrin in the control of gastrointestinal mucosal growth. Postulated roles of gastrin processing intermediates in the growth and differentiation of the gastric and colonic epithelium. Used with permission from *Am J Physiol* 277: pp. G6–11, 1999.

Negative regulation of gastrin release depends on the decreased pH levels following acid secretion, which is mediated by the paracrine effects of other hormones, such as somatostatin.

One of the most important functions of gastrin is to regulate gut mucosal growth and IEC proliferation, and it has been recognized as the single most important trophic hormone of the stomach [84]. The regulatory effect of gastrin on gut mucosal growth was initially identified more than three decades ago in two separate reports. Studies by Johnson et al. [101] and Crean et al. [102] revealed that administration of a synthetic gastrin analog, pentagastrin, increases protein synthesis and parietal cell mass in rats. These findings were quickly confirmed using the natural amidated gastrins, G-17 and G-34. G-17 and G-34 produced maximal stimulation on DNA synthesis in the oxyntic mucosa, duodenum, and colon at doses of 13.5 and 6.75 nmol/kg, respectively [84,103]. Removal of endogenous gastrin by antral resection causes mucosal atrophy that can be overcome by exogenous administration of gastrin. Subsequently, increased fundic mucosal proliferation was also demonstrated in rodent models following the administration of H_2-receptor antagonists, resulting in hypergastrenemia [84]. These animals exhibit the gastric gland elongation and increased cell proliferation. In humans, infusion of gastrin at high doses was shown to increase gut epithelial cell proliferation [104]. Consistently, patients with Zollinger–Ellison syndrome are associated with an increase in gastric mucosal growth due to high levels of gastrin in the circulation [27].

A number of gastrin gene transgenic or knock-out mouse strains have been developed, in which gastrin concentrations in circulation is increased or decreased dramatically [99,105]. Overexpression of either unprocessed gastrins or the amidated gastrins, such as G-17 and G-34, in the transgenic mice increases 2-fold elevation in serum-amidated gastrin and induces gut mucosal growth, particularly induction in numbers of parietal cells that is associated with an elevated gastric acid secretion [99,100]. These observations are further supported by results in athymic nude mice bearing xenografts of a transplanted human gastrinoma demonstrating gastric and duodenal mucosal hyperplasia. Interestingly, with aging, there is a progressive loss of parietal cells and expansion of the mucous neck cell proliferation [84]. This condition resembles the human condition of atrophic gastritis, which is characterized by a loss of gastric glands, progressive loss of parietal cells and increased plasma gastrin levels [84] (Figure 7]. On the other hand, gastrin-deficient mice exhibit a significant decrease in parietal cell mass (by 35%) compared with wild-type control animals [106,107]. Another study also show that in gastrin knock-out mice, ECL cells appear to be clustered toward the bottom of the gastric gland and there is a reduced rate of cell migration to the base of the gland (Figure 7]. As opposed to the results using amidated gastrin, experiments using G-gly show colonic proliferation, sparking renewed interest in a role for gastrin in precursor products in colonic growth. Mice overexpressing progastrin truncated at glycine-72 (MTI/G-GLY) exhibit elevated serum and mucosal levels of G-gly compared with wild-type mice. MTI/G-GLY mice display increases in colonic mucosal thickness (by 43%) and in the percentage of goblet cells per

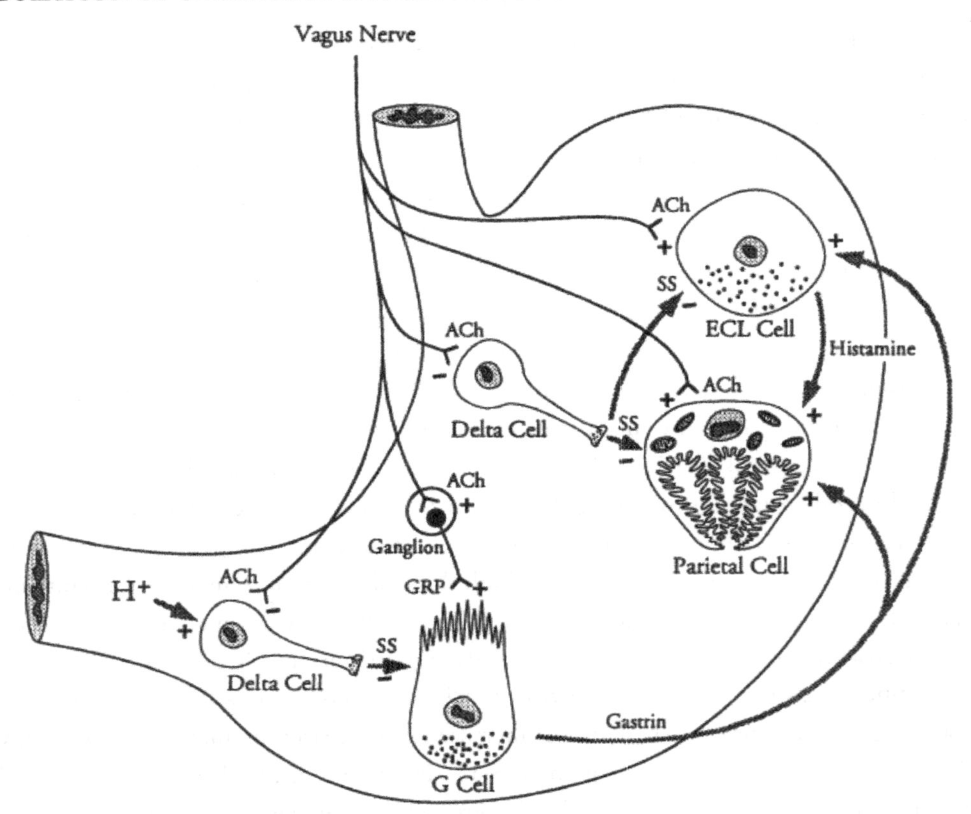

FIGURE 7: The interactions of the vagus affecting acid secretion and gastrin release. Vagal stimulation of the parietal cell occurs through M3 cholinergic receptors and via the release of histamine and gastrin from enterochromaffin-like (ECL) cells and G-cells, respectively. Somatostatin (SS) exerts a tonic basal inhibition of both the parietal cell and the G-cell. Vagal stimulation suppresses somatostatin release from delta cells, thereby disinhibiting these cells. Used with permission from *Yale J Biol Med* 67: pp. 145–51, 1994.

crypt (by 41%) [108]. Furthermore, administration of G-gly increases colonic mucosal thickness by ~10% and colonic proliferation by 81% in gastrin-deficient mice. These experiments using gastrin transgenic and knock-out models clearly show that gastrin is a potent stimulator of GI mucosal growth and epithelial cell proliferation.

In addition, gastrin receptor antagonists also have variable effects on gastrin-stimulated GI cancer growth [97]. For example, administration of proglumide (a potent gastrin-receptor blocker) inhibits the growth of MC-26 tumors *in vivo* and prolongs survival of tumor-bearing mice [99,109].

Hoosein et al. [110] demonstrate that gastrin stimulation is possibly attributable to an autocrine mechanism. Although discussion on the role of gastrin in the pathogenesis of GI cancers at detail is beyond the scope of this chapter, readers can be referred to understand most recent development of this area through several review articles [27,84,97,99,100,111].

CHOLECYSTOKININ

CCK is a member of the gut–brain family of peptide hormones, and it is produced by endocrine cells located predominantly in the proximal small intestine (duodenum and jejunum) as well as by the neurons in the myenteric plexus and brain. This gut peptide was first described in 1928 by Ivy and Oldberg [112] as a contaminant in impure secretin preparations. After complete purification and sequencing, CCK has been shown to be crucial for gallbladder contraction and pancreatic enzyme secretion. Other physiological functions of CCK in the GI tract include inhibition of gastric emptying, stimulation of bowel motility, potentiation of insulin secretion, and trophic effects on the pancreas and gut mucosa. Under biological conditions, CCK release is stimulated by fats, proteins, and amino acids.

Various forms of CCK are derived from the posttranslational modification of products of the pro-CCK gene which produces a cocktail of peptides with varying numbers of amino acids, each of which includes the minimal epitope for bioactivity [27,84]. In human embryonic development, CCK first appears in the duodenal and small intestinal mucosa at around 10 weeks of gestation, and the concentration progressively increases as gestation progresses and continue to present in endocrine cells of the GI tract. In the rat, maximal CCK concentration is present in the duodenum and jejunum. A high level of CCK is achieved in the suckling period, at around 14 days, and a decline is observed with weaning, and adult levels are reached at 28 days. CCK-58 is the largest form of the hormone, whereas CCK-8 is the smallest fragment containing 8-amino acids with complete biological activity [113]. The molecular forms of CCK are diverse and appear to be tissue-specific. For example, the most abundant form of CCK in the brain is CCK-8, although significant amounts of large carboxy-amidated forms such as CCK-33, CCK-58, and CCK-83 have been also isolated. CCK functions by directly interacting with the specific G-coupled CCK receptors (Figure 8). After binding to the receptor, a sequence of events collectively occurs, which culminates in an increase in concentration of intracellular calcium which in turn leads to degranulation of the pancreatic acinar cells and enzyme secretion. Additionally, CCK also stimulates the growth of the pancreas in experimental studies, although evidence for CCK regulating the growth of the GI mucosa is limited (Figure 8).

An increasing body of evidence demonstrates the trophic effects of CCK on the pancreas. Administration of CCK alone or combination with secretin causes marked increases in pancreatic weight, DNA, RNA, and protein content in rats [114]. The trophic effect of CCK on the

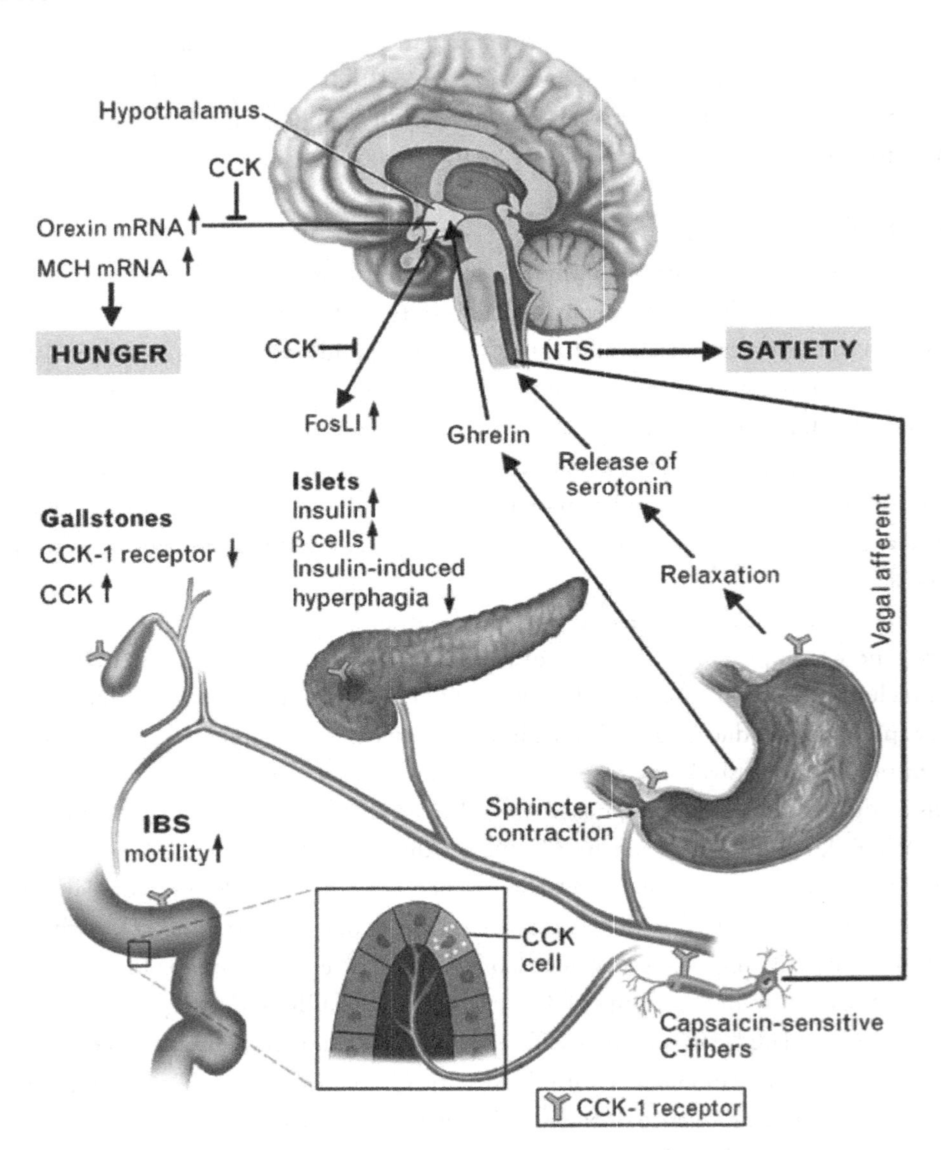

FIGURE 8: Cholecystokinin (CCK) is released by specific enteroendocrine cells following entry of food into the duodenum. CCK released into the blood or acting locally on enteric nerves exerts several effects that together coordinate important postprandial responses. CCK activates CCK-1 receptors located on the gastric fundus and pyloric sphincter, respectively, causing relaxation of the stomach and contraction of the sphincter, which together reduce the entry of food in the duodenum. CCK acts on the pancreas to stimulate both exocrine and endocrine secretion. Used with permission from *Curr Opin Endocrinol Diabet Obes* 14:63–7, 2007.

pancreas is physiologically significant since amino acids infused into the duodenum leads to a marked increase in pancreatic growth, and this stimulatory effect is prevented by the CCK-receptor antagonist such as CR 1409. Chronic camostate feeding increases pancreatic growth, which is associated with increased CCK plasma levels. The administration of exogenous CCK produces similar increase in pancreatic growth, but the combination of camostate and CCK-8 induces an additive stimulatory effect on the pancreas. Inactivation of CCK-receptors by CR 1409 completely abolishes the trophic effects of exogenous CCK-8 and inhibits the effects of chronic camostate feeding. In addition, treatment with CR 1409 alone also decreases pancreatic weight, DNA, and protein content. Johnson and Guthrie [115] have reported that administration of CCK-8 at very low doses induced a significant increase in pancreatic DNA synthesis, although it did not stimulate the mucosal growth of the oxyntic gland area or duodenum. Taken together, these findings clearly show that CCK is a potent stimulant for pancreatic growth.

On the other hand, evidence has been reported, showing the role of CCK in the regulation GI mucosal growth. Treatment with CCK and secretin is shown to prevent atrophy in the jejunum and ileum of dogs given total parenteral nutrition (TPN) as a sole nutrient source [84,116]. Administration of this peptide also increases galactose absorption, suggesting that CCK is a positive enterotrophic factor for the gut. Fine et al. [117], using intestinal bypass models, found that trophic response induced by CCK and secretin in the small bowel is the indirect result of increased pancreatobiliary secretion, as opposed to a direct stimulatory effect of these peptides on the gut mucosa. In cultured rabbit jejunum and ileum preparations, Stange et al. [118] further confirmed the lack of a direct effect of CCK on small-bowel growth. Furthermore, CCK has also been proposed as a major mediator of the satiety response, which leads to the cessation of feeding when food is placed in the stomach or intestine [119]. Intravenous CCK inhibits food intake in rats, although the mechanisms by which CCK controls the appetite are yet to be clearly identified. In addition to the involvement of CCK in the induction of satiety, this hormone is also involved in several disease conditions, such as diabetes mellitus, gall stone disease, irritable bowel syndrome (IBS), and inflammation (Figure 8) [84,120].

Several studies examined the cellular mechanisms by which CCK regulates pancreatic and gut growth. It has been reported that CCK activates the MAPK cascade, leading to the activation of ERK, JNK, and p38 MAPK in the pancreas [84,121]. Other signaling pathways that are also involved in CCK-induced mitogenesis and cellular proliferation include the PI3K-mTOR (mammalian target of rapamycin)-$p70^{S6K}$ and eIF4A pathways [121–123]. CCK stimulates the phosphorylation and activation of $p70^{S6K}$ in rat pancreatic acini; this activation is blocked by inhibiting mTOR and PI3K. In addition, the PI3K-mTOR pathway activates protein synthesis by phosphorylating the binding protein eIF4E, the translation initiation factor that binds to the 7-methyl guanosine cap at the 5' end of most eukaryotic mRNA molecules.

SECRETIN

Secretin was initially identified by Bayliss and Starling in 1902 [124]. In the past century, the research of secretin has gone by many milestones, which includes isolation, purification, structural characterization, and chemical synthesis of secretin, establishment of its hormonal status, identification of the specific receptor, cloning of secretin and its receptor genes, and identification of secretin-releasing peptides. Secretin has been identified as a hormone-regulating pancreatic exocrine secretion of fluid and bicarbonate, gastric acid secretion, and gastric motility [125,126].

There are a few observations showing the involvement of secretin in the regulation of GI mucosal growth so far. Generally, secretin is shown to inhibit the trophic action of gastrin but it has no a direct antitrophic activity in the GI mucosa [27]. Secretin inhibits the gastrin-mediated stimulation of DNA synthesis in the gastric oxyntic gland region, duodenum, and colon [127,128]. This inhibitory effect of secretin is independent of its ability to inhibit gastrin-stimulated acid secretion. It is likely that secretin indirectly regulates GI mucosal growth by blocking the trophic effect of gastrin.

SOMATOSTATIN

Somatostatin (SST) is a natural peptide hormone secreted in various parts of the human body including the GI tract [129]. SST was first identified by Brazeau et al. in 1973 [130] and it was originally described as a growth hormone-releasing inhibitory factor, containing 14-amino acids. The single SST gene is expressed in numerous endocrine cells in the GI system, including gastric mucosa and pancreas. In the intestinal mucosa, a 92-amino acid precursor molecule is processed to release a 28-amino acid peptide, of which 14 amino acids occupy the N-terminal position [131]. SST acts via five different but related receptor molecules belonging to the superfamily of G-protein-coupled receptors [132,133].

SST is a regulatory–inhibitory peptide and functions as the universal endocrine off-switch. SST represses the release of growth hormones and all known GI hormones, and it also inhibits gastric acid secretion and motility, intestinal absorption, and pancreatic bicarbonate and enzyme secretion, and selectively reduces splanchnic and portal blood flow [129]. Importantly, SST also inhibits the growth of the GI mucosa and normal pancreas, and this effect is mediated through either an indirect mechanism such as inhibition of other trophic hormones or a direct mechanism via interaction with the SST receptor subtype 2 [134]. Experiments *in vivo* revealed that the SST administration decreases DNA synthesis and reduces the number of parietal cells in gastric mucosa and exocrine pancreatic cells [27,135]. Furthermore, SST given together with gastrin reduces gastrin-stimulated mucosal growth of the stomach. On the other hand, in the duodenum and jejunum, the effects of SST are less consistent, with nocturnal SST producing a slight decrease in DNA synthesis, suggesting that SST inhibits cell division in the mucosa of normal GI tract and, furthermore, antagonizes the trophic activity of gastrin. Similarly, there are inhibitory effects on rat mucosal growth of duodenum using the SST analog, sandostatin. The notion that SST represses GI mucosal growth is further supported by studies demonstrating that the normal adaptive hyperplasia noted in rats after 40% small-bowel resection is abolished by administration of SST.

In the stomach, SST is produced by fundic and antral D cells which are closely associated with parietal cells, enterochromaffin-like (ECL) cells and gastrin G cells either directly via cytoplasmic processes (paracrine secretion) or indirectly via the circulation (endocrine secretion) [136]. This close anatomical relationship provides the morphological base for the tonic inhibitory effect of SST on gastric acid secretion directly by inhibiting parietal cells and indirectly by inhibiting the release of histamine from the ECL cells and gastrin from G cells (Figure 7). Piqueras and Martinez [137] demonstrated that SST modulates the gastrin-ECL cell/parietal cell axis through SST receptor subtype-2. A synthetic SST analog peptide, octreotide, is clinically used, and its pharmacological actions in the GI tract include the inhibition of release of gastrin, motilin, secretin and vasoactive intestinal polypeptides, reduction in blood flow to the gut mucosa, and repression of intestinal motility and carbohydrate absorption. Octreotide treatment for patients with acromegaly prevents hormone hypersecretion from the tumor, normalizing circulating growth hormone and IGF-1 levels, thus altering tumor growth [133,138,139].

There are considerable amounts of results showing the effects of SST on the normal pancreatic growth [84,140]. The administration of SST reduces pancreatic weight, DNA, RNA, and protein content, whereas SST depletion with cysteamine stimulates pancreatic growth. In addition, treatment with cysteamine augments the trophic effect of bombesin on the pancreas. Blocking endogenous SST may release these inhibitory constraints and allow for increased proliferation of the normal pancreas. Although the exact mechanism underlying the inhibitory effects of SST remains largely unknown, the SST receptor subtype-2 associates and stimulates tyrosine phosphatase Src homology 2-containing tyrosine phosphatase-1 activity, which in turn arrests cells in the G_0/G_1 phase of the cell cycle associated with up-regulation of the cyclin-dependent kinase inhibitor p27kip1 and an increase in hypophosphorylated retinoblastoma protein levels.

GHRELIN

Ghrelin is a relatively new member of the gut hormones and was first isolated from the rat and the human stomach in 1999 [141]. Ghrelin is a 28-amino acid peptide and serves as an endogenous ligand for growth hormone secretagogue receptor (GHSR). Together with the recently discovered 23-amino acid obestatin, it is derived from a prohormone precursor (proghrelin) by the posttranslational processing. Cells immunoreactive to ghrelin are widely distributed in the gastric mucosa in domestic and laboratory animals and in humans. The greatest expression of ghrelin is in the stomach, particularly in endocrine A-type cells in the oxyntic mucosa, and smaller amounts were in the small intestine and colon. Ghrelin stimulates the release of growth hormone from the pituitary both *in vitro* and *in vivo* and plays an important role in the regulation of food intake, energy homeostasis, gastric emptying, and acid secretion [142,143]. In contrast, obestatin seems to induce the opposite effects. The active form of ghrelin known as acyl ghrelin binds to and activates its receptor, GHSR-1a, and crosses the blood–brain barrier [144]. Peripheral and central administration of ghrelin in rats stimulates acute food intake [145], whereas the circulating levels of ghrelin increase with fasting but fall in response to meal intake, which is proportional to the calorie load of the meal. Ghrelin secretion is increased by acetylcholine and gastric inhibitory peptide, but decreased by CCK, SST, insulin and infection with *Helicobacter pylori* [146,147]. Taheri et al. [148] reported that lack of sleep is also associated with an increase in ghrelin levels and a decrease in leptin levels.

Ghrelin and ghrelin receptor expression are found in developing GI fetal and neonatal tissues, and substantial amounts of ghrelin are also identified in colostrum, suggesting that it has a potential role in perinatal development. Subcutaneous administration of ghrelin to pregnant rats is shown to increase body weight of newborn animals, which may be due to its possible role in stimulating the gut development in the postnatal period [149]. In another study, exogenous administration of ghrelin reduces the gastric growth in suckling rats, as indicated by a decrease in levels of gastric mucosa weight, DNA synthesis, and total DNA content [149]. Furthermore, treatment with ghrelin produces a significant reduction in the pancreatic weight and pancreatic amylase enzyme activity, and the inhibitory effects of ghrelin result probably from the hypothalamus immaturity in postnatal animals. Several studies by Kotunia et al. [150,151] show that repetitive intragastric administration of ghrelin in neonatal pigs fed with milk formula results in a significant depletion in body weight and reduces small intestine length, and it also causes a remarkable reduction in villous length and thickness of the tunica mucosa and tunica muscularis in the jejunum and ileum. In humans, the expression of ghrelin in the stomach is increased during infancy which may be related to the increase in gastric acid output, growth hormone secretion, and meal intake [143,144,149,150, 152–154]. Although intensive research has been carried out on ghrelin over the past decade, there remain many interesting questions regarding ghrelin-related biology. These include the identification of the pathways regulating ghrelin production and release from the GI tract, the enzyme that catalyzes its acyl-modification, and the continuing search for its physiological actions, especially its role and mechanism in modulating GI mucosal growth under various pathological condition.

BOMBESIN/GASTRIN-RELEASING PEPTIDE

Bombesin (BBS), a 14-amino acid peptide that was originally isolated from the extracts of skin from European amphibians in 1970, is analogous to mammalian gastrin-releasing peptide (GRP) [155]. BBS/GRP was found in a variety of species including rats, guinea pigs, dogs, and humans, and

BBS/GRP-like immunoreactivity is widely distributed throughout the GI tract, predominantly in the neuronal populations of the gut and in the acid- and gastrin-secreting portions of the stomach [156]. In general, BBS/GRP functions as a universal on-switch and plays predominantly stimulatory effects. BBS/GRP stimulates the release of almost all GI hormones, intestinal and pancreatic secretions, and motility [157]. Although the most important functions of BBS/GRP are to regulate antral gastrin release and gastric acid secretion, this peptide also stimulates growth of the GI mucosa and pancreas.

Increasing evidence indicates that BBS/GRP is a potent trophic factor in the GI tract and pancreas [158]. Since BBS/GRP stimulates the release of gastrin, it is obvious that its effects are opposite to those of somatostatin. It has been shown that the administration of BBS/GRP for 7 days at 8-h intervals stimulates DNA synthesis and total DNA and RNA content in the gastric and colonic mucosa [157], but this stimulatory effect is attenuated by SST. Another experiment found that removal of the gastrin-secreting cells by antrectomy and subsequent administration of CCK receptor inhibitor prevents the proliferative effects of BBS/GRP, suggesting that the mucosal growth activities of BBS/GRP are mediated through gastrin and CCK stimulation [159]. Lehy et al. [160] observed that BBS/GRP administration twice daily for 1 week induces the gastrin-cell proliferation and increases the antral gastrin content. Gastric weight, fundic and antral mucosal weight, and the number of parietal cells are also increased in neonatal rats treated with BBS/GRP.

BBS/GRP is also shown to stimulate growth of the small-bowel mucosa in rats fed with liquid elemental diet [84]. When BBS/GRP was administered for 11 days, it not only prevents jejunal mucosal atrophy but also enhances ileal mucosal growth as determined by mucosal weight, RNA, DNA, and protein content. Furthermore, Chu et al. [161] showed that BBS/GRP-mediated stimulation of small intestinal mucosal growth is regulated by factors that are independent of luminal contents and pancreaticobiliary secretion. Several studies also revealed that BBS/GRP stimulates colonic mucosal growth [115] as well as the growth of the pancreas [160]. In addition, BBS/GRP improves integrity of the gut mucosal epithelium after exposure to severe burn injury by decreasing burn-induced gut mucosal atrophy and epithelial cell apoptosis [157,162]. These findings suggest that BBS/GRP plays an important role in intrinsic gastric mucosal defense system against various luminal noxious substances. Qiao et al. [163] recently reported that silencing BBS/GRP receptors suppresses tumor growth and reduces metastatic potential of neuroblastoma *in vitro* as well as *in vivo*.

OTHER GI HORMONES

As mentioned earlier, there are more than 50 gut hormones and peptides synthesized and released from the GI tract, of which only small proportions have been vigorously investigated for their potential roles in the regulation of GI mucosal growth. The glucagon-like peptides, GLP-1 and GLP-2, are peptide hormones released from the gut endocrine L-type cells in response to the products of a mixed carbohydrate and fat meal [164]. L-type cells are most abundant in the ileum and colon in human intestine, and these cells are the second most numerous populations of endocrine cells after ECL cells [165]. GLP-1 is shown to increase pancreatic islet mass by stimulating β-cell proliferation; it also promotes differentiation of exocrine cells or immature islet progenitors. On the other hand, GLP-2 stimulates cell proliferation in GI mucosa, leading to an expansion of the GI mucosal epithelium [166]. Mice treated with GLP-2 exhibit elongated villi by enhancing crypt cell proliferation and decreasing enterocyte apoptosis [167]. Furthermore, exogenous GLP-2 enhances mucosal regeneration in pathological conditions, such as colitis and small-bowel enteritis [84].

Vasoactive intestinal peptide (VIP) is a 28-amino acid peptide and also plays a role in regulating gut mucosal growth through a way similar to secretin [168]. Although VIP given alone shows no significant effect on DNA synthesis or total DNA content in the GI mucosa, VIP given together with pentagastrin prevents the gastrin-induced trophic effects on gastric or colonic mucosae [169].

In addition, neurotensin (NT) is produced by the endocrine N-type cells of the jejunum and ileal mucosas, its major functions in the GI tract are to stimulate pancreatic and biliary secretions and to suppress small-bowel gastric motility [170]. There are considerable amount of results showing that NT also promotes growth of the gastric antrum, small bowel, colon, and pancreas [84,170].

Gastric-inhibitory polypeptide (GIP) is a 42-amino acid peptide that is secreted by specific K-type endocrine cells. GIP is also implicated in the regulation of gut mucosal growth [171]. Additional information about the involvement of other gut peptides in controlling growth of gut mucosa and pancreas is available in several review articles [84,149,172,173].

• • • •

Peptide Growth Factors in GI Mucosal Growth

The intestinal tissues express a variety of peptide growth factors that modulate several functional properties of different intestinal cell populations, including the intestinal epithelium and lamina propria cell populations. These peptide growth factors are characterized by relatively low molecular weight of less than 25 kDa, and they generally exert their effects through binding to specific high-affinity cell-surface receptors present on their respective target cells [95,97,174,175]. In contrast to classical peptide hormones that are released to circulatory system for delivery to distant target organs or cells, peptide growth factors tend to act locally on adjacent cells (paracrine or juxtacrine action) or on the same cells that have expressed the peptide factors (autocrine action) (Figure 9). Although the full variety of peptide growth factors that are implicated in the control of the intestinal epithelium and nonepithelial compartment of the intestine remains to be demonstrated, an increasing body of evidence shows the diversity of these peptides and their importance in the regulation of GI mucosal growth. These peptides include members of epidermal growth factor (EGF) family, the transforming growth factor-β (TGF-β) family, fibroblast growth factor (FGF) family, the insulin-like growth factor (IGF) family, the trefoil factor family (TFF), and few other peptides described in Table 1. These growth factors are generally produced by intestinal mesenchymal tissues and regulate epithelium and nonepithelial tissue functions, such as cellular proliferation, differentiation, migration, and cytoprotection.

As shown in Table 1, the constituents of this network possess multiple functional properties and exhibit pleiotropism in their cellular sources and targets; they are highly redundant in several dimensions. For example, each cell type appears to produce more than one peptide growth factors, whereas each peptide may be produced by multiple different cell populations within the intestinal tract. In addition, most cell populations express receptors specific for more than one peptide growth factors, while receptors for a single growth factor are present on multiple cell types. Furthermore, functional effects of a certain growth factor are modulated by the co-presence of other factors, and structurally related multiple members of a peptide growth factor may interact with a single receptor.

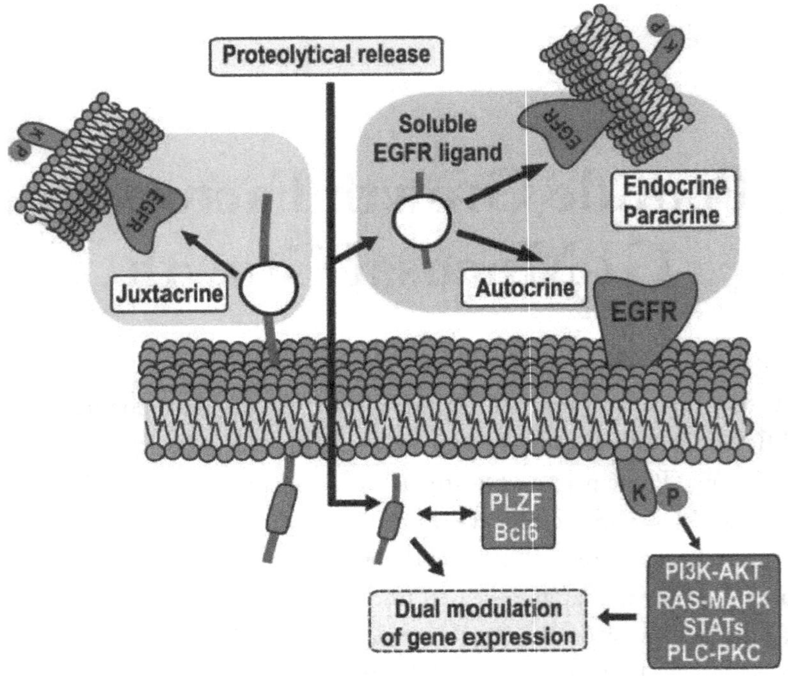

FIGURE 9: Ectodomain shedding of EGFR ligands and its consequences for signaling. Membrane-bound molecules can activate the EGFR of neighbor cells (juxtacrine mechanism). Following proteolytic release, the soluble EGF module activates the EGFR of neighbor cells (paracrine), distant cells (endocrine), or at the own cell membrane (autocrine). Ectodomain shedding also results in a free cytoplasmic tail, which can interact with other cytoplasmic proteins to modulate gene expression. Used with permission from *J Cell Physiol* 218: pp. 460–6, 2009.

In this chapter, we briefly describe the roles and mechanisms of peptide growth factors of EGF family, TGF-β family, IGF family, and FGF family in the regulation of GI mucosal growth.

EGF FAMILY

The EGF family consists of different peptides including EGF, TGF-α, amphiregulin, heparin-binding EGF (HB-EGF), betacellulin, epiregulin, and neuregulin. All of them exhibit mitogenic activity upon binding to four different high-affinity receptors: EGFR/ErbB1, HER2/ErbB2, HER3/ErbB3, and HER4/ErbB4. Members of the EGF family are characterized by three properties: (i) the ability to bind to the EGF receptors, (ii) the capacity to mimic the biological activities of EGF, and (iii) amino acid sequence similarity to EGF. In fact, the overall identity of sequences within all members of the EGF family is approximately 20% [176]. EGF and TGF-α are the

TABLE 1: Peptide growth factors and their target cells in the gastrointestinal system

GROWTH FACTOR	GASTROINTESTINAL TARGET CELLS
TGF-α/EGF	Epithelial cells
	Endothelial cells
TGF-β	Epithelial cells
	Endothelial cells
	Immunocytes
	Mesenchymal cells
IGF	Epithelial cells
	Endothelial cells
	Fibroblasts
CSF	Immunocytes
	Hematopoietic stem cells
	Epithelial cells
FGF	Epithelial cells
	Endothelial cells
	Mesenchymal cells
TFF	Epithelial cells
	Goblet cells
HGF	Smooth muscle cells
	Epithelial cells
	Endothelial cells

TGF-α, transforming growth factor alpha; EGF, epidermal growth factor; TGF-β, transforming growth factor beta; IGF, insulin-like growth factor; CSF, colony-stimulating factor; FGF, Fibroblast growth factor; TFF, trefoil factor; HGF, hepatocyte growth factor.

prototype members of the EGF family. In the GI tract, EGF is produced in submaxillary glands and Brunner's glands in the duodenum. Small amounts of EGF are also produced within exocrine pancreas; it is also present in gastric juice and within the intestinal lumen. In the early postnatal life, the breast milk is also a major source of EGF [177]. EGF produces a variety of biological responses, most of which are involved in the regulation of cell proliferation and/or differentiation, cell movement, and survival in epidermal as well as epithelial tissues [177,178]. The EGFR/ErbB1 is well characterized and possibly the most biologically important receptor for EGF family members in the GI epithelium, although additional EGFRs are also identified in normal and fetal GI tissues. In the GI epithelium, EGF promotes development of the intestinal mucosa, promotes cell proliferation and differentiation, and also enhances mucosal healing after injury [27,179].

EGF is a potent stimulator of cell division in epithelial and nonepithelial cell types in the GI tract, and specific EGF receptors are widely distributed in many cell types. The widespread distribution of EGFRs, including a variety of cells committed to terminal differentiation, suggests that EGF and/or another member of the EGF family have a range of biological functions beyond their mitogenic activity. EGF is shown to modulate the expression of enzymes involved in the production of cellular polyamines, to up-regulate intestinal electrolyte and nutrient transport in the enterocyte, to stimulate expression of brush border enzymes, to attenuate intestinal damage, and to enhance GI mucosal healing after injury [180,181]. For example, Berlanga-Acosta et al. [182] reported that the continuously infusing EGF for long periods (14 days) by implanting osmatic minipump subcutaneously in rats increases intestinal epithelial cell proliferation and induces the crypt and villus areas in the small intestine. This stimulatory effect of EGF occurs as early as 24 h after EGF infusion, but its maximal stimulation is observed 6 days thereafter. Another study conducted by Cellini et al. [183] revealed that infusing EGF directly into intra-amniotic fluid in pregnant rats during the last 8 days of gestational period results in a significant increase in fetal weight, intestinal villus height, and DNA synthesis within the crypts.

Recently, Kang et al. [184] demonstrated that oral administration of recombinant EGF together with probiotic bacteria for 14 days stimulates intestinal development and reduces the incidence of pathogen infection and diarrhea in pigs. Intestinal length, jejunal and duodenal villus heights are greater in animals treated with EGF and probiotic bacteria compared to controls and animals treated with EGF or probiotic bacteria alone. Immunohistochemistry with antibodies against proliferating cell nuclear antigen (PCNA) revealed that the proliferation of intestinal cells was significantly greater in the EGF+ probiotic bacteria administered group. Studies conducted by these authors in their earlier observations [186] also showed that the administration of recombinant EGF increased mean villus height, crypt depth, and enterocyte proliferation compared to control mice fed with phosphate-buffered saline. EGF also has a beneficial effect on the intestinal development and growth of newly weaned mice. Based on these observations, it has been suggested that the combination of EGF with probiotic approach could provide the possibility for formulating dietary supplements for children during their weaning transition stages [184,185].

TGF-α has been characterized as a product of many cell types, including most epithelial cells in the GI tract. The mRNA and protein levels of TGF-α have been identified in human and rodent stomach, small intestinal and colonic epithelium. The biological activities of TGF-α are mediated through the same receptor as EGF. In 1999, Montaner et al. [188] showed the immunolocalization of TGF-α in the rat gastroduodenal region. In the stomach, the surface and gastric pit cells showed increased immunostaining of TGF-α in the cytoplasm and basolateral and apical membranes. In the duodenum, the enterocytes co-express both TGF-α and EGFR in the supranuclear area. These immunolocalization studies demonstrate that the co-expression of TGF-α and EGFR in the rat GI tract suggests a functional role in the establishment and maintenance of the epithelial renewal. Simultaneously, *in vivo* studies conducted in rats showed that TGF-α administration resulted in a significant increase in mucosal weight, DNA and protein content, and villus height in jejunum and ileum and also induced crypt depth in jejunum and ileum. Important functions of TGF-α in the GI tract include trophic effects on mucosa, stimulator of epithelial and nonepithelial cell proliferation, alterations of expression involved in the mucosal development, promotion of growth of intestinal neoplasia, enhancement of epithelial restitution, and stimulation of angiogenesis.

The biological actions of EGF and its family of peptides are mediated via interaction with EGFR, which is detected throughout the fetal and neonatal GI tract. EGFR is predominantly expressed in the villus tip cells in young pigs still fed on maternal milk, but after weaning, it was more concentrated in Brunner gland and in goblet cells [189]. Kuwada et al. [190] reported that total cellular EGFR protein and mRNA transcript levels are relatively unchanged during cell differentiation *in vitro*, but the expression of surface EGFR and patterns of expressed EGF-ligand changed significantly. It is likely that EGFR system is regulated during intestinal epithelial cell differentiation primarily at the level of ligand expression. In addition, it has been shown that integrin α5/β1 mediates fibronectin-induced epithelial cell proliferation through the activation of the EGFR.

In another study conducted by Duh et al. [191], they have demonstrated the specific roles of EGFR during embryonic gut development by using EGFR knock-out mouse model. EGFR activation appears to accelerate the maturation rate of goblet cells and to induce differential crypt/villus proliferation pattern in early embryonic mouse gut. Moreover, the human milk induces fetal small intestinal cell proliferation through the mechanism involving different tyrosine kinase signaling pathways via the EGFR [192] (Figure 9). Taylor et al. [193] showed that the activation of EGFR enhances intestinal adaptation after massive small-bowel resection as indicated by taller villi, deeper crypt areas, and augmented enterocyte proliferation. Defective EGFR signaling in mutant mice exhibits increased apoptosis and reduction in bcl-2 family gene expression [194,195].

TGF-β FAMILY

TGF-β is a family of structurally homologous dimeric proteins consisting of at least three isoforms, TGF-β1, TGF-β2, and TGF-β3 [196]. The prototypic member of the TGF-β family in the GI tract is TGF-β1, although other two isoforms of TGF-β may be also detected in all GI tract tissues and accessory organs. TGF-β is synthesized as a large precursor propeptide [197]. Despite intracellular cleavage, the TGF-β1 dimer remains in a biological inactive complex with the two propeptide segments through noncovalent association, the so-called latent form. The biological processes regulating the bioactivation of TGF-β from its latent state have not been completely defined. TGF-β1 has been found to bind to several specific cell surface TGF-β receptors (TβRs) localized in responsive cells. There are five different types of receptors (TβRI through TβRV), and among them, TβRI and II isoforms are Ser/Thr-specific protein kinases that are believed to be primarily responsible for TGF-β induced cellular responses in the GI tract [198]. It has been shown that TβRI and TβRII work in a cooperative fashion: ligand binding to the TβRI facilitates activation of the associated TβRII, which then activates the intracellular signaling machine via Smad proteins [199]. Within the small intestine, TGF-β expression has been found in lamina propria and almost all the epithelial cells [200]. The major activity of TGF-β is to inhibit the growth of most cell types, including epithelial and endothelial cells, but in some instances, TGF-β also stimulates the growth of certain mesenchyme cells, such as in skin fibroblasts [196]. The GI mucosal growth inhibitory responses to TGF-β have been intensively investigated *in vitro* as well as *in vivo*, although the mechanism underlying the inhibitory effects of TGF-β remains to be fully understood.

Targeted disruption of the TGF-β gene by gene knock-out technology results in multiple focal inflammatory cell infiltration and/or necrosis, indicating the role of TGF-β in both inflammation and tissue repair. Mice homozygous for the mutated TGF-β allele exhibit no gross developmental abnormalities at birth, but they develop severe and multifocal inflammatory diseases that affect several organs, including diffuse inflammation in the stomach and intestine. In fact, TGF-β deficiency leads to severe pathology, causing death at about 20 days of age associated with dysfunction of the immune and inflammatory system, showing its essential role as a potent regulator of the immune system. Increased expression of TGF-β is also found in the GI mucosa after acute epithelial injury and in patients with active inflammatory bowel diseases.

Results from our laboratory and others show that increasing the levels of TGF-β inhibits intestinal epithelial cell proliferation through activating TβRI/Smad signaling cascade following polyamine depletion (Figure 10) [201–203]. The addition of TGF-β to the culture medium significantly decreased the rate of DNA synthesis and final cell number. Increased activation of endogenous TGF-β/TβRI signaling in polyamine-deficient cells is also associated with an inhibition of intestinal epithelial cell growth, which is partially prevented by the addition of immunoneutralizing anti-TGF-β antibody or inactivation of TβRI activity [202,203]. We have further demonstrated that

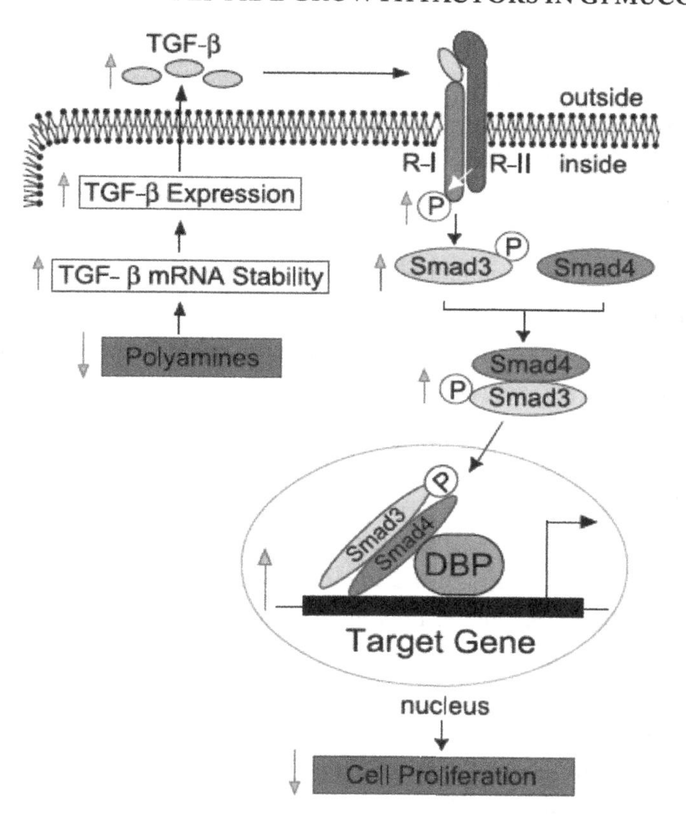

FIGURE 10: Schematic diagram depicting the role of TGF-β/Smad signaling pathway in the inhibition of normal intestinal cell proliferation following polyamine depletion. In this model, polyamines are the negative regulators for expression of the TGF-β gene, whereas Smad proteins are the downstream intracellular effectors of activated TGF-β receptors. Decreased cellular polyamines, by either inhibition of their synthesis, stimulation of the catabolism, or suppression of polyamine uptake, increase expression of TGF-β through stabilization of TGF-β mRNA, enhance the release of TGF-β, and subsequently phosphorylate (P) the TGF-β type II receptor (R-II). The phosphorylated R-II activates TGF-β type I receptor (R-I), induces the formation of Smad3/Smad4 heteromeric complexes, and stimulates their nuclear translocation. The activated Smads in the nucleus bind to the specific DNA site and cooperate with Smad DNA-binding partners (DBP) such as some activating protein-1 proteins to activate or repress transcription of specific target genes, thus leading to the inhibition of normal intestinal epithelial cell proliferation following polyamine depletion. Used with permission from *Am J Physiol* 285: pp. G1056–67, 2003.

Smad proteins are the immediate downstream effectors of activated TGF-β/TβRI signaling since Smad silencing prevents inhibitory effects of exogenous TGF-β treatment or activated endogenous TGF-β/TβRI pathway via polyamine depletion on intestinal epithelial cell proliferation [201].

Gebhardt et al. reported that TGF-β acts as a novel potent inhibitor of human intestinal mast cells [204]. In this study, mast cells were isolated from the human intestinal mucosa, purified, and cultured in the presence of stem cell factor (SCF) with or without TGF-β1. TGF-β1 was found to dose-dependently inhibit SCF-dependent growth of human intestinal mast cells by decreasing proliferation and enhancing apoptosis. In another study, the prenatal porcine intestine is shown to have low levels of endogenous TGF-β ligand and receptor density, which is associated with an induction in trophic response to enteral diets [205]. There is also a reporter showing that in fetal pigs, the TGF-β ligands are predominantly localized to the crypt epithelium, but staining intensity increased markedly just before term and shifted to the villus epithelium in newborn pigs [206].

In addition, TGF-β and gastrin-releasing peptides (GRP) jointly regulate intestinal epithelial cell division and differentiation [207–210]. The treatment with TGF-β together with GRP is found to inhibit intestinal epithelial cell growth and to induce apoptosis much higher than those observed in cells exposed to TGF-β or GRP alone. This combined treatment also induces an induction in cyclooxygenase-2 expression and prostaglandin E2 production through activating p38MAPK pathway in cells stably transfected with GRP receptor. In another study, TGF-β transcriptional activity was found to be upregulated in the small intestine after infection of mice with a parasite *Trichinella spiralis*, which leads to small intestinal inflammation [211]. The TGF-β signaling pathway also plays an essential role in intestinal stem cell development and organogenesis [207,212,213].

IGF FAMILY

IGF family is constituted by two ligands, IGF-I and IGF-II, which are single chain polypeptides consisting of 70 and 67 amino acids, respectively [214]. IGFs are secreted as small peptides (7.5 kDa) that are structurally related to insulin and display multiform effects on cell growth and metabolism in the GI tract [215,216]. Both IGF-I and IGF-II exert their mitogenic activities through interaction with specific IGF-receptors (mainly IGFR-I and IGFR-II), and they are capable of modulating epithelial cell kinetics by stimulating proliferation and inhibiting apoptosis [214]. In rodents, IGFI and IGFII are expressed in diverse sites with intrinsic biological activities. IGF-I is produced by intestinal mesenchymal cells in rats, and it is shown to increase proliferation of intestinal epithelial cells [217]. IGF-II is expressed at high levels in the fetus and lower levels at the adult stages, and its synthesis has been observed in adult liver and extrahepatic tissues [218].

The stimulatory effects of IGFs on the intestine had been identified over almost two decades ago. Recently, therapeutic indications are also defined for a range of candidate bowel disorders and diseases in which accelerated intestinal repair is desirable. IGFs are the potent stimulator of cell proliferation in the intestinal crypts, spurring progression through G1- to the S-phase of the cell cycle. Exogenous administrations of IGFs pharmacologically and systemically to neonatal or adult rats increase intestinal mucosal growth [219]. Continuous administration of IGF-I to adult rats for 14 days causes preferential growth of the GI organs, increase gut weight as a fraction of body weight by up to ~32%, which is accompanied by increases in crypt cell population and villus cell density. Consistently, overexpression of the IGF transgenically increases growth at the intestinal muscle layers [220]. IGFs have been also demonstrated to promote wound healing in the GI mucosa to exert trophic effects within the intestine and to enhance tumor growth through autocrine mechanism. There is also evidence that IGFs are important in the pathogenesis of fibrosis in Crohn's disease.

Experiments of IGF-I administration *in vivo* have shown both linear and cross-sectional growth of the GI organs affecting the mucosal and muscularis layers proportionally [221]. These findings suggest clinical application in bowel conditions characterized by impaired growth and repair processes [214,222–224]. It is likely that bowel resection, chemotherapy-induced intestinal mucositis, radiation enteritis, and the inflammatory bowel diseases are candidate target conditions that may be beneficial from IGF administration in the first instance [225].

FGF FAMILY

Members of huge FGF family are 16–18 kDa proteins that control normal growth and differentiation of mesenchymal, epithelial, and neuroectodermal cell types [226]. FGFs also play key roles in growth and survival of stem cells during embryogenesis, tissue regeneration, and carcinogenesis. Both acidic and basic FGFs are the best-characterized members of the FGF family. Although there are limited results available on the expression and physiological functions of FGFs and their receptors in the GI tract, several studies have described the presence of FGF family peptides and their specific receptors in the intestine [174,227]. It has been shown that FGFs appear to act as autocrine growth factors and stimulate intestinal mucosal growth and epithelial cell division. In cultured IEC-6 cells, administration of FGFs increases cell proliferation; and this effect is further enhanced by the addition of heparin. This treatment mimics the *in vivo* situation of growth factors binding to the extracellular matrix.

Furthermore, FGFs are also shown to promote intestinal epithelial restitution after wounding through TGF-β-dependent pathway *in vitro* [228]. *Helicobacter pylori* are the major pathogen for peptic ulcers and chronic atrophic gastritis, and they are also implicated in the pathogenesis of gastric cancer [229]. FGF2 is one of the pro-angiogenic factors and shown to enhance healing of gastric mucosal damage associated with *Helicobacter pylori* infection. CagA protein and peptide glycan of *Helicobacter pylori* are phosphorylated by SRC family protein kinases to activate SHP2 phosphatase when they are injected to human gastric epithelial cells [229]. Since SHP2 is a component of docking protein complex for FGF signaling, SHP2 activation results in FGF signaling activation.

FGF signaling pathway networks with Wnt signaling pathway during a variety of cellular processes including early embryogenesis and gastrointestinal morphogenesis [174,230]. Expression analyses on different FGFs revealed that FGF-18 and FGF-20 isoforms are predominantly expressed in epithelial cells derived from the GI tract, and they seem to be the direct targets of the canonical Wnt signaling pathway. Wnt signals are transduced through Frizzled seven-transmembrane-type receptors and activate the β-catenin pathway. Wnt-induced transcriptional complex activates transcription of target genes. It has been reported that the promoter of FGF-18, FGF-20, or FGF-7 contains several TCF/β-catenin binding sites, but the exact role of Wnt signal in the regulation of FGF gene transcription remains to be fully defined.

OTHER FACTORS

Several other peptide growth factors and cytokines have also been found to play an important role in maintaining mucosal cell growth under biological and various pathological conditions. These factors include Trefoil factor (TFF) family [231,232], hepatocyte growth factor (HGF) [233], and colony-stimulating factors (CSF) [234]. All TFF, HGF, and CSF are shown to stimulate GI mucosal growth, promote wound healing, modulate epithelial cell apoptosis, and protect the epithelial integrity from damage in response to stressful environments.

• • • •

Luminal Nutrients and Microbes in Gut Mucosal Growth

The GI epithelium is critically located at the interface between the body and environment. This interface acts as a peace keeping force positioned between two opposing armies, and it must placate the local environment to prevent conflict from erupting. As befitting a role requiring a high degree of tact and dexterity, the GI epithelium incorporates an array of strategies to facilitate peaceful communication between luminal contents, including nutrients and microbes and the mucosal renewal system, thereby preserving its tissue homeostasis. During starvation, the small intestinal mucosa atrophies rapidly, with a reduction in cell proliferation being noted within hours of food withdrawal [235]. This inhibition in the GI mucosal growth occurs even when the overall nutritional state of the animals are maintained by total parenteral nutrition (TPN).

There is an increasing body of evidence showing that luminal nutrients stimulate gut mucosal growth by local direct effect at their site of absorption [27,236,237], and this direct action does not result from the use of the nutrients as sources of energy by mucosal cells, since nonmetabolizable absorbed substrates, such as galactose and 3-*O*-methyl-D-glucose, also promote mucosal cell proliferation [238]. Although the exact mechanism underlying this process remains unclear, it is likely that the workload of absorption determines the gut mucosal growth response. Luminal nutrient also stimulates intestinal mucosal growth indirectly by releasing gut hormones from the distal small intestine, colon and pancreas [239]. In addition, adaptive changes in small intestinal mucosal mass are generally associated with parallel changes in segmental absorptive function [236], but the magnitude of induction of individual transport processes can be selectively affected by the specific nature of the nutrients within the lumen. In rats receiving TPN, for example, infusion of D-glucose into the small intestine increases specific transport capacity for glucose, while substituting protein isocalorically for carbohydrate in the diet increases amino acid transport capacity and reduces the transport capacity for galactose. Recently, interaction between animal and bacterial cells is also shown to play an important role in the regulation of GI mucosal growth [240,241].

LUMINAL FACTORS

Luminal factors include a variety of nutrients, secretions, and other essential components in the diet or produced in the lumen of the GI tract that have been known to function physiologically to stimulate gut mucosal growth (Figure 11). A large body of evidence has accumulated and strongly suggests that luminal factors are the principal stimulus for GI growth [27,242,243]. Initially, Dowling [244] proposed "luminal nutrition" as the underlying regulator of GI mucosal growth which is derived by the absorbing enterocytes from absorbed nutrients. Ricken and Menge [245] later defined it as "the presence of nutrient material in the lumen" and proposed that "this does not imply that nutrients must be absorbed to have effects." A later definition of luminal nutrition includes direct effects of nutrients, luminal growth factors, and the releasing trophic hormones by the ingested food substances. Almost two decades ago, an excellent review published by Johnson [27] on the regulation of GI mucosal growth emphasized the roles of luminal nutrition in the regulation of gut mucosal growth under physiological conditions and he also named these luminal nutrients and growth factors as "local nutrition" to avoid confusion among scientific community.

Stimulation of gut growth by luminal nutrients is verified by infusing a wide variety of nutrient substances into Thiry–Vella (T–V) loops. Jacobs et al. [246] showed that infusion of liquid elemental diets causes hyperplasia of bypassed intestinal mucosa. Clarke [247] proposed the interesting concept of work-load model, based on the experiments in which the morphology and crypt cell metaphase accumulation rates were examined in starved rats with a self-emptying loop that

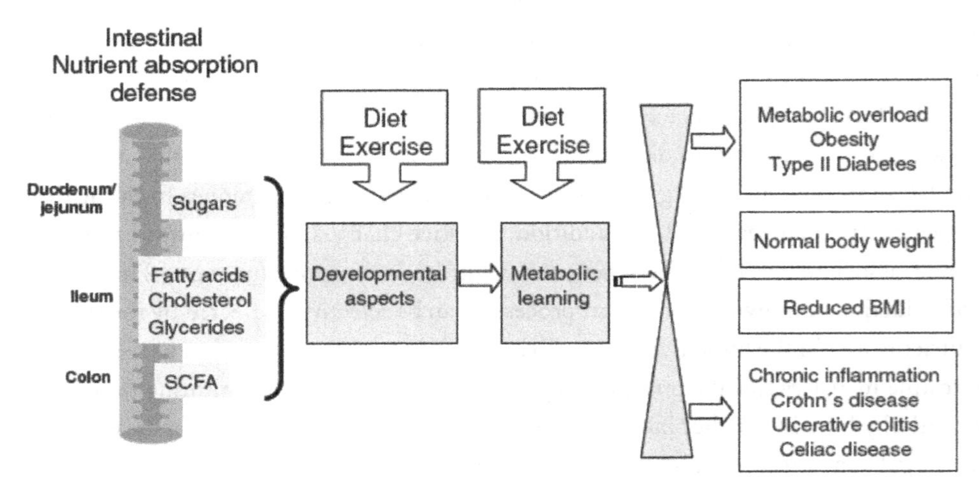

FIGURE 11: Nutrient adaptive responses of the intestinal mucosa. Adaptation to nutrition and luminal factors by developmental aspects and metabolic learning. Used with permission from *Horm Metab Res* 38: pp. 452–4, 2006.

was designed to test when the group of animals had infused with distilled water into the loop and to determine whether or not a noncaloric luminal stimulus would affect the parameters of intestinal growth. He also tested with different isotonic solutions, such as galactose, methylglucoside, or sodium chloride infusion, to stimulate cell production to the same extent as glucose. Philpott et al. [248] examined the role of luminal versus systemic factors in promoting intestinal recovery using the refeeding of previously malnourished infant rabbits using T–V loops. They found that the luminal factors stimulate intestinal repair during the refeeding of malnourished infant rabbits. Studies in the pig as well as in rat have shown the similar results as luminal nutrients stimulate the mucosal cell growth indirectly by releasing enterotrophic hormones from intestine [240,249,250].

The relative potency of various sugar compounds in stimulating mucosal growth had been intensively investigated by several laboratories utilizing parenteral nutrition. The infusion of dextrose intragastrically or into the mid ileal region causes enormous gut hyperplasia [251,252]. There is a linear correlation between the amount of dextrose infused into the gut and mucosal growth. Infusion of 5% glucose, galactose or fructose solution did not elevate intestinal mass compared to saline infusion alone. However, 5% sucrose, maltose, or lactose solution significantly increases intestinal mass content, suggesting that disaccharides are more trophic to the intestinal mucosa than monosaccharides [252]. These trophic responses were abolished when hydrolysis of disaccharides was prevented. Mannitol, a nonabsorbable sugar, is ineffective in promoting growth, while the infusion of nonmetabolizable sugar 3-*O*-methyl glucose does promote intestinal growth [238]. These findings indicate that the functional work load of the absorbing epithelium, including work of hydrolysis, plays a crucial role in the stimulating effect of nutrients on GI mucosal growth. Another experiment determined how luminal nutrients control the GI mucosal growth by using various supplemental diets. Mice raised on a high-carbohydrate diet exhibited almost 35% greater intestinal mass, primarily in the proximal intestine, compared to those raised on a carbohydrate-free diet. High carbohydrate fed mice had elevated levels of sugar uptake, but switching those mice to carbohydrate-free diet led to a rapid decrease in glucose uptake, which suggests that there is no preprogramming of intestinal function [252]. Furthermore, different hydrolyzable disaccharides are shown to have similar stimulatory effects on small intestinal mucosal mass when fed with a mixture of amino acids [236]. Dietary supplements with ornithine led to an induction in small intestinal mucosal growth, probably due to an increased polyamine production, but other individual amino acids stimulate mucosal cell proliferation through different mechanisms. Ingested food is the major source of polyamines in the lumen of the upper small bowel, and shortly after meal, polyamine concentrations in the duodenal and jejunal lumen reach its peak and it occurs as early as 120 min after meal luminal polyamine content and then gradually returns to the fasting level. Our laboratory investigations show that luminal polyamine content plays a major role in the GI mucosal growth, and we emphasize more on the luminal polyamines in the next chapter.

Epithelial cells of the small intestine are nevertheless capable of incorporating orally administered amino acids into protein [253]. Hirschfield and Kern [254] suggested that luminally derived amino acids are important in the nutrition of the small intestinal mucosa during protein deprivation. Studies in parenterally nourished rats have shown that infusion of relatively low concentrations of amino acids (<5%) stimulates intestinal mucosal growth to a greater degree than isotonic saline or isocaloric dextrose. This process was further delineated when individual amino acids were infused and each amino acid had its specific trophic potential. It appears from the various studies that histidine is a better stimulator of GI mucosal growth than valine or glycine in small intestine [236]. Attention has been recently focused on the role of the amino acid glutamine in small intestinal growth, and we discuss the importance of glutamine in the following section in this chapter.

Dietary lipid content in a mixed nutrition has a moderate trophic effect on the small intestinal mucosal growth. Intragastric infusion of long-chain triglycerides was found to enhance the adaptive response to partial small intestinal resection in the rat compared to carbohydrate, protein, or medium-chain triglycerides administration (Figure 11) [255]. Further studies showed that long-chain free fatty acids have a greater effect than long-chain triglycerides. The nature of ingested fat also influences the absorptive function of the small intestine, perhaps as a result of changes in the nature of lipid incorporated into the cell membrane of the enterocytes. It has been noticed that essential fatty acids have an important role in mucosal proliferative responses, as deficiency of this dietary component attenuates the adaptive response of the small intestinal mucosa [256]. In addition to fat substances, dietary fiber also plays a prominent role in the regulation of small intestinal growth and function. The addition of nonabsorbable kaolin has no trophic effect on the small intestinal hyperplasia, although it increases glucose and water absorption *in vitro*. Supplementing an elemental diet with α-cellulose increases small intestinal weight and cell proliferation compared with the same elemental diet alone. Dietary fiber can be fermented by lower gut bacteria to release short-chain fatty acids (SCFAs). Goodlad et al. [257] reported that small intestinal and colonic mucosal growth is significantly increased by the most fermentable fibers and that these enterotrophic effects are abolished in germ-free animals.

Various GI secretions, including pancreatic and biliary secretions, are also shown to act as stimulants for the intestinal mucosal growth [258]. Feeding stimulates the stomach, liver, pancreas, and small intestine to secrete the compounds that are able to promote intestinal mucosal growth and are implicated in the pathogenesis of intestinal mucosal hyperplasia. Luminal nutrients also stimulate structural and functional regeneration in the intestine through a process involving IGF-1 and glucagon-like peptide 2 (GLP-2). Nelson et al. [259] investigated the relationship between IGF-1 and GLP-2 responses and mucosal function in rats fasted for 2 days and then refed for 2 or 4 h by continuous intravenous or intragastric infusion or *ad libitum* feeding. Fasting induced a significant decrease in plasma IGF-1 and GLP-2 levels, body weight, intestinal protein, DNA content, and villous height, but these changes are attenuated by exogenous IGF-1, GLP-2, or refeeding. Parenteral nutrition and the absence of luminal feeding result in impaired intestinal growth and differentiation of enterocytes. In a separate study, administration of GLP-2 is also shown to have trophic effects on the intestine [249,260]. The specific roles of these GI secretions, including gastrin, secretin, CCK, SST, and various growth factors, have been discussed in the previous chapters and have also been discussed in several excellent reviews [84,214,232].

MICROBES IN HEALTH AND MUCOSAL GROWTH

The GI mucosa is in continuous contact with prokaryotic symboints. Until recently, it has been recognized that microbes present in the lumen of gut affect GI health and functions including the regulation of the GI mucosal growth [241,261]. The epithelial cells lining the intestine function to keep bacteria from invading the body, but they also have mutually beneficial relationship with these intestinal flora, collectively termed as "microbiota" or "microflora," which regulate a wide variety of physiological functions of the gut (Figure 12). Prebiotics are nondigestive foods able to selectively stimulate the growth and/or activity of a limited number of colonic bacteria, where probiotics are

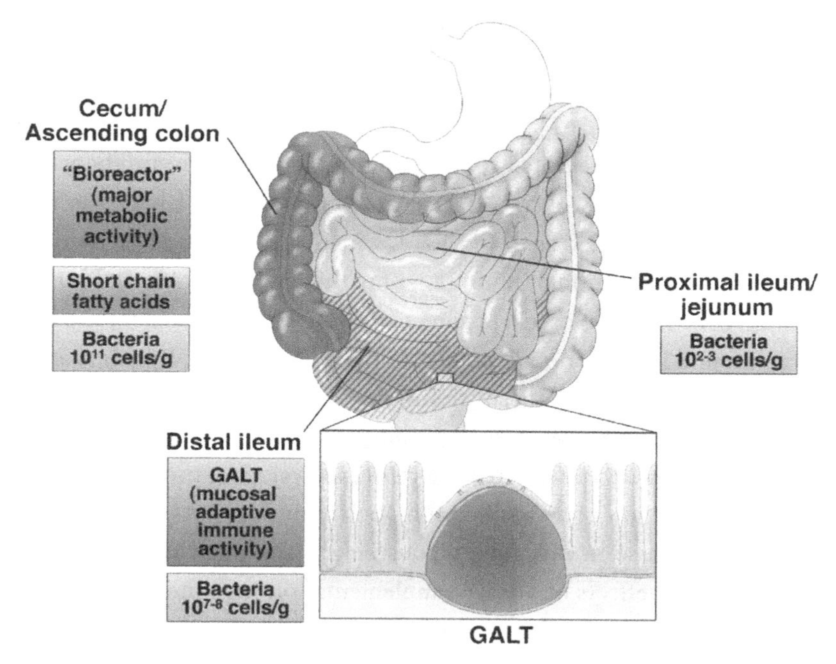

FIGURE 12: Preferred sites of commensal/probiotic interaction with the gut. Cecum/ascending colon is a "bioreactor" with the greatest amounts of bacteria, metabolic activity, and short chain fatty acids (SCFA) fermentation. Concentration of SCFA diminishes along the colon. The distal ileum is enriched in GALT (Peyer's patches) and is the dominant site of luminal sampling and mucosal adaptive immune activity. Used with permission from *Gastroenterology* 136: pp. 65–80, 2009.

defined as living microorganisms which when administered in adequate amounts confer a health benefit on the host [241,261], including gastric bile and pancreatic secretions, attach to epithelial cells and colonize in the intestine, and also they provide the same beneficial functions and activities that have evolved from the normal intestinal microbiota [262].

Several recent studies have enhanced our ability to understand the interactions between the host and its intestinal microflora and the importance of microflora in maintaining intestinal homeostasis [237,262,263]. Among the substrates considered as prebiotics are the oligosaccharides, inulin, fructo-oligosaccharides, galacto-oligosaccharides, and lactulose. Studies showed that prebiotics have beneficial effects on various markers of health [261,264]. Roy et al. [265] described that the dietary carbohydrates escaping digestion/absorption in the small bowel and prebiotics undergo fermentation of these substances in the colon, giving rise to SCFAs and use them as a major source of energy for colonocytes (Figure 12). Thus, the dietary supplementation with bacterial fermentation substrates, usually complex carbohydrates, can increase luminal concentration of SCFAs. SCFA, in general, and butyrate, in particular, enhance the growth of *lactobacilli* and *bifidobacteria* and play a major role in the physiology and metabolism of colon [262,266]. The effects of prebiotics on cell proliferation, differentiation, apoptosis, immune function, mineral absorption, and GI peptides synthesis have been extensively studied recently [261,263,264,267,268]. Currently, the food industry is also making efforts to commercialize prebiotics and exploit a wide array of application for their potential health benefits [261,263,267].

Several studies also have attempted to identify specific positive health benefits of probiotics using different bacterial strains. Beneficial effects exerted by probiotic bacteria in the treatment of human diseases are broadly classified as those effects which arise due to the activity in the large intestine and are related to colonization of inhibition of pathogen growth [269]. Human milk (colostrum or mature milk) constitutes an excellent source of commensal bacteria for the infant gut [266]. Among the bacteria found in human milk belong to the species *Staphylococcus*, *Lactococcus*, *Enterococcus*, and *Lactobacillus*, and some of these strains are considered as potentially probiotic species [267,270]. The health promoting properties of probiotics are known to be strain-dependent. Thus, strain identification and characterization are important in developing probiotics for human use. For example, the oral administration of a specific probiotic strain *Pediococcus acidilactici* (Pa) to piglets exhibited a remarkable increase in intestinal villous height and crypt depth [240]. In this study, authors also investigated the effects of dietary supplementation with the probiotic Pa on the piglet intestine, circulating lymphocytes, and aspects of piglet performance during the first 42 days after weaning. Pa supplementation positively influenced weight and post-weaning average daily weight gain of treated piglets, associated with the larger number of proliferating enterocytes than in control animals. These studies showed that the probiotics supplementation is able to protect the piglet's small intestinal mucosa, improving local resistance to infections in the stressful weaning period.

Another study conducted recently by Awad et al. [271] showed the beneficial effect of probiotic strain *Lactobacillus* species on the growth of small intestinal mucosal architecture in broiler chickens. The body weight and average daily weight gain are significantly increased by the dietary inclusion of *Lactobacillus* species, which is associated with increased glucose transport, intestinal villous height and crypt depth ratio compared with control group. Results also showed the improvement of intestinal architecture and epithelial nutrient absorption after administration of probiotic strain *Lactobacillus* species. Probiotic supplementation not only improves the growth of intestinal mucosa but also augments intestinal host defense by regulating apoptosis and promoting cytoprotective responses [272]. Lin et al. [273] demonstrated that administration of *Lactobacillus rhamnosus* GG (LGG) reduces chemically induced intestinal epithelial apoptosis *in vitro* and *ex vivo* as measured by staining for apoptotic markers in murine models and that LGG also prevents necrotizing enterocolitis (NEC) in preterm infants. Soluble proteins produced by probiotic bacteria LGG are shown to activate Akt, inhibit cytokine-induced epithelial cell apoptosis, and promote cell growth in human and mouse colon epithelial cells and in cultured mouse colon explants [273]. These findings also suggest that probiotic bacterial components are useful for preventing cytokine-mediated GI diseases. Probiotic approaches are and will be confounded by the diversity of the human microbiota and its plasticity in the face of the varied human diets and genetic backgrounds. Based on the data from both *in vitro* and *in vivo* models, probiotics offer great potential benefits and might be used to treat intestinal functional and inflammatory disorders, metabolic syndromes as well as intestinal nociception [261,264,267,268,274,275].

DIETARY SUPPLEMENTS

The GI tract is the only part of the body that directly comes in contact with a wide variety of nutrient molecules before they are absorbed. Various nutritional supplements, such as vitamins, amino acids, nucleic acids, and sugars, are shown to have trophic effects on GI mucosal growth, epithelial cell proliferation, and differentiation [276]. It has been widely accepted that the amino acid glutamine plays an important role in the regulation of GI mucosal metabolic functions and growth [277–279]. Several studies showed that intestinal glutamine metabolism not only acts as a nutritionally important portion of the energy generation, but also as the precursor or key factor of a number of important metabolic pathways of other amino acids, especially those leading to the synthesis of ornithine, citrulline, arginine, and proline. Among all amino acids, glutamine metabolism has had a greater influence on a variety of aspects of clinical nutrition [277]. There is evidence showing that the removal of glutamine by starvation of cultured intestinal mucosal cells inhibits cell proliferation [277], whereas the glutamine supplementation in rats exhibits a significant increase in the villous heights of small intestine. In contrast, the numbers of villi per unit length of bowel decrease when glutamine synthesis is inhibited by methionine sulphoximine or in the animals fed with glutamine-free diet [280]. In this study, it has been also noticed that there is a breakdown of the epithelial junctions in the glutamine-deprived and glutamine synthetase-inhibited intestines. These findings support the notion that glutamine in diet supplements is necessary for normal intestinal mucosal growth and for maintenance of the intestinal mucosal integrity. Wiren et al. [281] demonstrated the importance of glutamine in the regulation of cell differentiation by using cultured Caco-2 and HT-29 cells.

However, there are some controversial evidences about the role of glutamine supplements in the GI mucosal growth. It has been reported that oral glutamine supplementation alone does not sufficiently induce protein synthesis in the jejunal mucosa of malnourished rats, regardless of the total food intake or the presence or absence of glutamine supplementation [282]. Recently, Shyntum et al. [283] reported that dietary sulfur amino acid supplementation stimulates ileal mucosal growth rate after massive small-bowel resection in rats. In this study, they observed that resected and sulfur amino acid supplemented rats exhibited increased ileal adaptation as indicated by the increase in full-thickness wet weight, DNA, protein content, and mucosal crypt depth and villous height. Despite this negative observation, in particular experimental condition, the emerging results suggest that glutamine supplements are important for maintaining normal mucosal architecture including tight junctions.

Other important diet supplements are water- and fat-soluble vitamins which are essential for intestinal epithelial cell growth and epithelial cell proliferation [284,285]. Among them, vitamin A and its bioactive metabolites (retinoic acid) are well characterized as important agents that modulate a variety of GI physiological functions, including mucosal growth and cell proliferation. Uni et al. [286] found that vitamin A deficiency interferes with the normal growth rate in chickens. Decreased vitamin A alters the functionality of the small intestine by reducing epithelial cell proliferation and maturation. Similar findings were also observed in rats and showed that vitamin A deficiency modifies the maturation and differentiation processes of the small intestinal mucosa [287]. In a study conducted in calves that were supplemented with vitamin A, villous heights in the ileum and villous height to crypt depth ratios in the jejunum are enhanced in comparison with those from control animals. In addition, vitamin D, vitamin C, and Riboflavin in diet supplements also have similar beneficial effects on GI mucosal growth and epithelial tissue homeostasis [284–288].

Several studies also show the importance of dietary nucleotides in the regulation of GI mucosal growth and development [289,290]. Exogenous supplementation of nucleotides was found to have a beneficial effect on intestinal growth after parenteral supplementation. The wet weight of the jejunal mucosa and content of total protein and DNA are higher in the rats supplemented with a mixture of nucleotide and nucleosides than those observed in the rats fed with parenteral solution alone. The morphometric analysis showed that there is a significant increase in the villous height after nucleic acid supplementation, supporting the notion that dietary nucleotides are also crucial for normal mucosal cell growth and functions.

Sodium butyrate supplementation is also shown to enhance the GI mucosal growth and improve several indices of gastrointestinal biological functions in piglets after weaning [291]. In addition, various minerals, such as zinc, magnesium, and potassium, in diet supplements also affect physiological functions of the intestine and show the beneficial effects on intestinal mucosal growth [292–294].

• • • •

Polyamines in the Regulation of Mucosal Growth

The natural polyamines spermidine and spermine and their diamine precursor putrescine are ubiquitous small basic molecules that are found in all eukaryotic cells and are implicated in many aspects of cellular physiology [295,296]. Although the concentration of intracellular polyamines is in the millimolar range, free polyamines are considerably less abundant, as they are bound to various cellular anions including DNA, RNA, proteins, and phospholipids [295,297]. Polyamines are essential for mammalian cell growth and development, and gene disruptions of ODC or S-adenosylmethionine decarboxylase, two key rate-limiting enzymes in polyamine biosynthesis, are lethal at early stages of embryonic development [297,298]. Our previous studies [299–303] and others [295,304] showed that normal intestinal mucosal growth depends on the supply of polyamines to the dividing cells in the crypts and that reductions in cellular polyamines inhibit intestinal epithelial cell (IEC) proliferation *in vivo* as well as *in vitro*.

Polyamines also regulate IEC apoptosis, which occurs in the crypt area to counterbalance the continual proliferation of intestinal epithelial stem cells, and at the luminal surface of the colon and villous tips in the small intestine, where differentiated IECs are lost [296,299,305]. Although the specific molecular processes controlled by polyamines remains to be fully defined, increasing evidence indicates that polyamines regulate intestinal epithelial integrity by modulating the expression of growth-related genes [299,306]. Polyamines positively regulate the transcription of growth-promoting genes such as *c-fos*, *c-jun*, and *c-myc* [299,307] and negatively affect the expression of growth-inhibiting genes including p53 [301,308], JunD [306,309], and TGFβ/TGFβ receptors [201,203,310] through modulations at the posttranscriptional level. In this chapter, we will give an overview of the roles of polyamines in GI mucosal growth and highlight the changes in cellular signals that are activated following increased or decreased levels of cellular polyamines, particularly focusing on the implication of polyamines in the regulation of gene transcription, mRNA stability, and translation.

POLYAMINE METABOLISM

The biosynthesis and the catabolism of the polyamines are carefully controlled processes in all eukaryotic cell types. Cellular polyamines are highly regulated by their biosynthesis, degradation, and transport. Polyamine metabolism involves both forward and reverse component pathways, although the cellular control of the regulatory enzymes and the polyamine transporter act in concert to maintain appropriate levels of the individual polyamines. Polyamines are derived from two amino acids, ornithine and methionine. Polyamine biosynthesis depends on the activation or inhibition of ornithine decarboxylase (ODC), which catalyzes the first rate-limiting step in polyamine synthesis

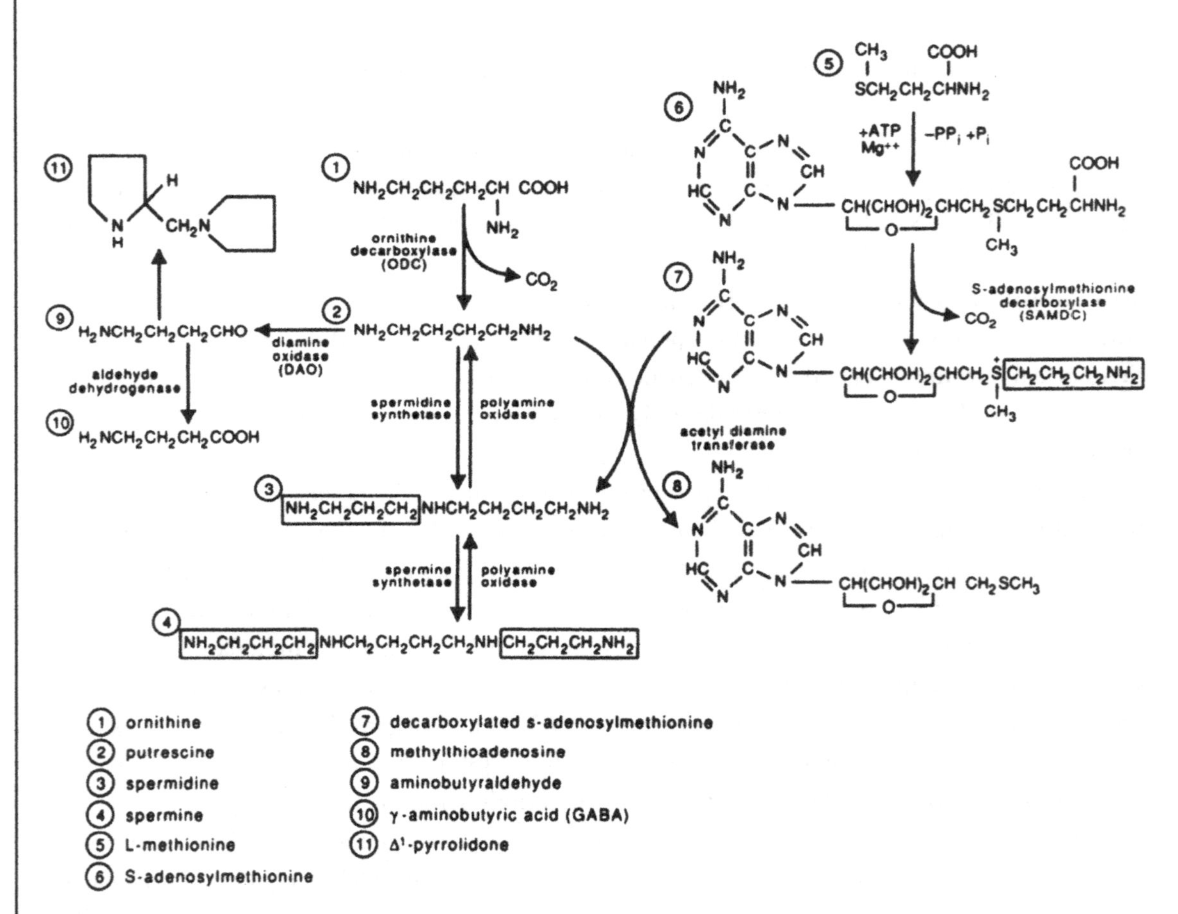

FIGURE 13: The biosynthesis of the major physiological polyamines and their precursor putrescine. Putrescine has 4 CH groups, spermidine 7, spermine 10. They carry 2, 3, and 4 positive changes that reflect the strength of their binding ability. Enzyme involved in the reactions are numbered. Used with permission from *Am J Physiol* 1991; 260:G795–806.

(Figure 13). ODC decarboxylates the amino acid ornithine to form putrescine; propylamine groups are then added to one or both amino groups of putrescine to form spermidine and spermine [295]. Methionine is the precursor for S-adenosylmethionine (AdoMet). The decarboxylation product of AdoMet is the precursor of the aminopropyl moieties of spermidine and spermine.

On the other hand, polyamines are degraded by diamine oxidase and spermidine/spermine-*N*-acetyltransferase [295,297]. Putrescine, spermidine, and spermine are interconverted according to cellular physiological needs [295,298]. Polyamine homeostasis is regulated by feedback mechanism mostly by the polyamines themselves via the regulation of *de novo* synthesis, release, uptake, and catabolism. In mammalian cells, the degradation of ODC is facilitated by a specific ODC-antizyme [311,312], a protein that also appears to downregulate polyamine transport. DL-α-difluoromethylornithine (DFMO) is an irreversible inactivator of the ODC enzyme that specifically inhibits the ODC activity (Figure 13). The discovery of DFMO has provided an enormous stimulus to the field of mammalian polyamine biology, which unraveled various mechanisms related to polyamines over the past decades.

POLYAMINES STIMULATE MUCOSAL GROWTH BY ENHANCING GENE TRANSCRIPTION

Over the last decades, considerable progress has been made in understanding the roles of polyamines in the regulation of gene transcription, particularly early primary response genes, during the process of mucosal growth and wound healing in the GI tract. Most of these early primary response genes belong to the family of protooncogenes and are responsible for control of the cell cycle [313,314]. Because the expression of these early primary response genes is rapid and transient following injury or when normal quiescent cells are exposed to mitogenic substances, they have been thought to act as mediators linking short-term signals, immediately after cell surface stimulation, to proliferation by regulating the activation of specific genes. These early primary response genes, such as protooncogenes, code for sequence-specific DNA binding nuclear proteins with a potential to influence directly the expression of specific genes at the transcriptional level. Transcriptional mechanisms could be due to alterations in the intracellular location, activity, or generation of transactivating factors as well as numerous intracellular mediators that directly or indirectly bind *cis*-acting elements in the promoters of protooncogenes and influence the rate of transcription either positively or negatively. Therefore, activation of early primary response gene expression plays important roles in normal GI mucosal growth and healing after wounding.

The regulation of protooncogene transcription in regenerative tissues unquestionably is dependent on cell type and stimuli. We have gathered considerable evidence indicating that intracellular polyamines are involved in the regulation of transcription of protooncogenes during epithelial renewal in the GI mucosa. At physiological pH, polyamines, putrescine, spermidine, and spermine possess two, three, and four positive charges, respectively. Together with magnesium ions, they account for majority of the intracellular cationic charges. In contrast to magnesium, intracellular polyamine levels are highly regulated by the cell according to the state of growth. Polyamines are believed to exert their effects by binding to negatively charged macromolecules, such as DNA, RNA, and proteins, influencing the chromatin structure and sequence-specific DNA–protein interactions, which alter the level of gene transcription. It has been shown that the entry of cells into the cell cycle is accompanied by large increases in newly synthesized polyamines by ODC, which occurs in concert with increases in the expression of a number of early response genes [307,315]. Polyamine synthesis usually precedes DNA synthesis, and the depletion of intracellular polyamines by treatment with DFMO results in a decrease in cell proliferation [295,299,301].

Polyamines Regulate Epithelial Renewal by Altering Expression of Protooncogenes

A series of observations from our previous studies has demonstrated that polyamines, either synthesized endogenously or supplied luminally, are absolutely required for the process of cell division during healing of gastric and duodenal mucosal stress ulcers [302,316] and for normal intestinal mucosal growth [302,303]. Stress significantly increased ODC activity, which remained markedly elevated over that of corresponding controls for 12 h after the period of stress in both gastric and duodenal mucosa. Increases in mucosal content of putrescine, spermidine, and spermine paralleled the changes in ODC activity and peaked 4 h after stress. Administration of DFMO not only inhibited the increases of ODC activity and polyamine levels but also almost totally prevented healing in both tissues. In addition, oral administration of polyamines immediately after stress increased the normal rate of healing and prevented the inhibition of repair caused by DFMO [303]. Spermidine or spermine accelerated healing better than putrescine. The delayed recovery of DNA synthesis and contents of DNA and RNA after stress in the DFMO-treated animals was also significantly prevented by exogenous polyamines.

We have further demonstrated that increased expression of *c-fos* and *c-myc* protooncogenes in regenerative gastric mucosa after stress is regulated by cellular polyamines [316]. Exposure of rats to stress results in a rapid increase in the activity of ODC, which is associated with an increase in *c-myc* gene expression in the gastric mucosa. The significant increase in ODC activity occurred at 4 h of stress and peaked 4 h after a 6-h period of stress. Baseline expression of *c-Myc* mRNA and protein was enhanced dramatically after 6 h of stress and remained significantly elevated for 8 h. By 12 h, after the period of stress, the expression of *c-myc* has returned to near normal levels. Administration of DFMO [500 mg/kg) in stressed animals not only prevented the marked increases in ODC and polyamines, including putrescine, spermidine, and spermine, but also inhibited the induced expression of the *c-myc* gene in the gastric mucosa. The *c-myc* mRNA and protein levels were decreased by ~70% immediately after the 6-h stress period and totally absent following a 4-h recovery period from the 6-h stress in DFMO-treated rats.

In cultured intestinal epithelial cells, we also demonstrated that polyamines stimulated normal cell growth in association with their ability to regulate expression of the protooncogenes [307,314]. Depletion of cellular polyamines by DFMO prevented the increased expression of *c-myc* and *c-jun* during log-phase growth and also significantly reduced steady-state levels of *c-myc* and *c-jun* mRNAs during the plateau phase. Treatment with DFMO also totally prevented the increased expression of *c-fos* when 5% dialyzed fetal bovine serum was given after 24-h serum deprivation. The remarkable parallelism that exists between increased intracellular polyamines and induced expression of protooncogenes during healing or normal cell growth led us to test the possibility that polyamines are involved in the regulation of transcription of the protooncogenes in intestinal epithelial cells.

Polyamines are Required for Protooncogene Transcription

In order to determine the role of intracellular polyamines in the regulation of protooncogene transcription, we carried out three experiments in IEC-6 cells, a line derived from rat small intestinal crypt cells [301,317]. In the first experiment, depletion of cellular and nuclear polyamines by treatment with DFMO for 4 or 6 days significantly decreased the steady-state levels of *c-myc* and *c-jun* mRNA. The changes in *c-myc* and *c-jun* transcription paralleled those of their respective cytoplasmic mRNA levels. The rate of *c-myc* transcription decreased by 55% on day 4 and by 60% on day 6 in DFMO-treated cells. The *c-jun* transcription in DFMO-treated cells decreased by 75% on day 4 and 85% on day 6. These low rates of *c-myc* and *c-jun* transcription in cells treated with DFMO returned toward normal levels after administration of exogenous spermidine (5 μM). The transcription of *c-myc* and *c-jun* in cells grown in the presence of DFMO plus spermidine for 4 and 6 days was indistinguishable from that of control cells. Because the decreased rate of transcription of *c-myc* and *c-jun* in the DFMO-treated cells is completely overcome by exogenous spermidine, inhibition of transcription of *c-myc* and *c-jun* mRNAs in the DFMO-treated cells must be related to polyamine depletion rather than secondary to an effect of DFMO unrelated to the inhibition of polyamine biosynthesis.

In the second experiment, we determined whether polyamine depletion prevented the increased transcription of *c-myc* and *c-jun* after exposure of quiescent cells to 5% dFBS, a known stimulator of these two genes [299,315]. IEC-6 cells were grown in the presence or absence of DFMO for 4 days, and serum was deprived for 24 h before experiments. Transcription rates of *c-myc* and *c-jun* were measured 3 h after administration of 5% dFBS. Results showed that 5% dFBS stimulated transcription rates significantly in normal quiescent cells. The increased rates of *c-myc* and *c-jun* were 8 and 3.5 times the control levels, respectively. These increases were prevented significantly by polyamine depletion. In polyamine-deficient cells, the rate of *c-myc* transcription slightly increased after exposure to 5% dFBS and was twice the control level. The increased transcription of *c-jun* by 5% dFBS was completely prevented in the DFMO-treated cells.

In the third experiment, we examined the effect of addition of spermidine to nuclei isolated from control and DFMO-treated cells on the transcription rates of *c-myc* and *c-jun* [315,318]. Cells were initially treated with or without DFMO for 6 days, and then the nuclei were isolated. When the transcription rates of *c-myc* and *c-jun* were examined before and after spermidine addition to isolated nuclei, significant differences between control and DFMO-treated cells were observed. There was no significant alteration in the transcription rate of *c-myc* or *c-jun* gene in the nuclei from control IEC-6 cells (without DFMO) after the addition of spermidine at different concentrations. In contrast, spermidine addition to polyamine-deficient nuclei from DFMO-treated cells resulted in a marked increase in the transcription rates of *c-myc* and *c-jun* genes. Administration of spermidine resulted in a 2- to 2.5-fold increase in *c-myc* and *c-jun* transcription without altering the transcription of GAPDH gene. The increase in *c-myc* and *c-jun* transcription in response to spermidine addition to nuclei was concentration dependent with a maximum increase observed at a concentration of 0.5 mM. The effect of spermidine addition on the transcription rates of *c-myc* and *c-jun* in isolated nuclei from DFMO-treated cells was not accounted for by replacement of positive changes, since increasing $MgCl_2$ concentration by 0.5 to 2.0 mM over the standard conditions of 5 mM $MgCl_2$ in the assay system had no effect on the transcription of any specific gene. These results are consistent with the data from other investigators [296], who have demonstrated that spermidine induces an increase in protooncogenes and other transcription factors in nuclei from certain types of cells.

To define the downstream targets of polyamine-induced c-Myc during GI mucosal growth and wound healing after injury, we have demonstrated that overexpression of the ODC gene increases c-Myc expression and inhibits p21Cip1 transcription as indicated by repression of p21Cip1-promoter activity and reduction of p21Cip1 protein levels [315]. In contrast, depletion of cellular polyamines decreases c-Myc but increases p21Cip1 transcription (Figure 14). Ectopic expression of wild-type *c-myc* not only inhibits basal levels of p21Cip1 transcription in control cells but also prevents increased p21Cip1 in polyamine-deficient cells. Experiments using different p21Cip1-promoter mutants further show that transcriptional repression of p21Cip1 by c-Myc following induction in cellular polyamines is mediated through Miz-1- and Sp1-binding sites within the proximal region of the p21Cip1-promoter in normal intestinal epithelial cells. These findings confirm that p21Cip1 is one of the direct mediators of induced c-Myc following increased polyamines

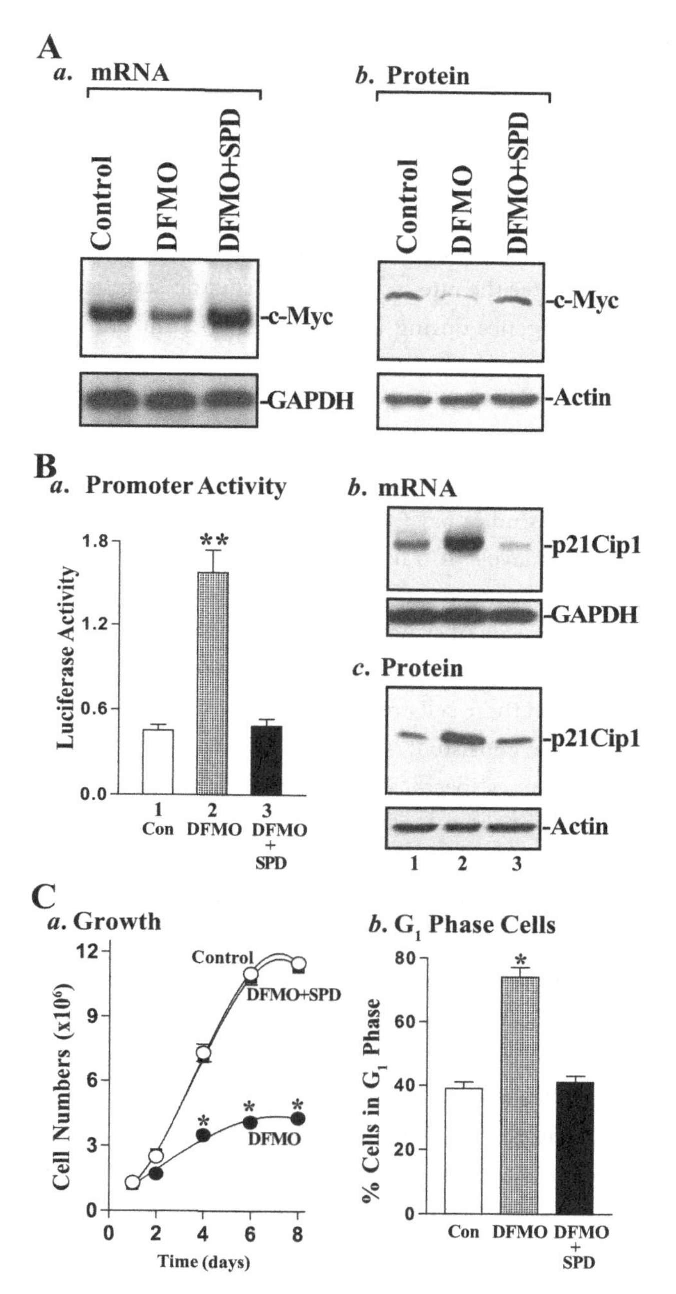

FIGURE 14: Effect of depletion of cellular polyamines by inhibiting ODC on c-Myc and p21Cip1 expression and intestinal epithelial cell growth. (A) Levels of c-Myc mRNA and protein. (B) p21Cip1 promoter activity and levels of p21Cip1 mRNA and protein. Cells were grown in medium containing either α-difluoromethylornithine (DFMO) alone or DFMO plus spermidine (SPD) and then transfected using the p21Cip1 promoter luciferase reporter construct. (C) Changes in cell growth as indicated by cell numbers and percentages of G1 phase cells. Used with permission from *Biochem J* 398: pp. 257–67, 2006.

and that p21Cip1 repression by c-Myc is implicated in stimulation of normal intestinal epithelial cell proliferation.

Taken together, these data clearly show that intracellular polyamines play a critical role in the regulation of transcription of protooncogenes in intestinal epithelial cells. In stress-induced gastric ulcers, mucosal polyamine levels dramatically increased in response to damage [302,316]. Because increased cellular polyamines alter the rate of protooncogene transcription, it is likely that increases in expression of the protooncogenes during healing of gastric mucosal stress ulcers are primarily caused by the activation of gene transcription.

Possible Mechanisms of Action of the Polyamines

The molecular mechanisms by which polyamines modulate transcription of protooncogenes in the GI mucosa are unclear. Several studies have suggested that changes in chromatin structure by polyamines may be correlated with levels of transcription of *c-myc* and other genes [313]. In cell-free systems, polyamines have been shown to affect B-to-Z conformational changes in DNA as well as changes in chromatin and nucleosomal structure [319]. Because the addition of similar concentrations of simple divalent cations such as Mg^{2+} has no specific effect on DNA structure, the alterations achieved by polyamines in these cell-free systems are not the result of simply altering the ionic environment. These findings are consistent with our previous results [299,301] and those of others showing that exogenous polyamines specifically reverse the inhibition of cell proliferation following polyamine depletion.

In addition to effects on DNA structure, polyamines have been shown to alter sequence-specific DNA–protein binding activities, which may affect the regulation of initiation, elongation, and termination during transcription. Intracellular polyamine levels may provide an ionic environment that could be modulated to alter the binding or release of transcriptional regulatory factors [320,321]. Some of these effects are polyamine specific, while others are due to the general cationic nature of these compounds [295,321]. Panagiotidis et al. [322] reported that at physiological concentrations polyamines specifically enhanced the binding of several proteins including upstream stimulatory factor (USF), transcription factor E_3 (TFE_3), immunoglobulin/enhancer binding protein (Ig/EBP), nuclear factor-interleukin-6 (NF-IL6), and Yin-Yang binding protein-1 (YY1) to DNA, but inhibited others such as octamer-binding protein-1 (Oct-1). Polyamines facilitate formation of complexes involving binding of more than one protein on a DNA fragment but do not influence DNA–protein contacts. The decreased rate of *c-myc* and *c-jun* transcription that we observed in polyamine-deficient IEC-6 cells may result from abnormalities in the interaction between transcription factors and their cognate DNA sequences. Clearly, further study is necessary to examine whether polyamines play a specific role in the interaction or synthesis of key transcriptional regulatory factors in intestinal epithelial cells.

INDUCED mRNA STABILIZATION AND GROWTH ARREST AFTER POLYAMINE DEPLETION

Negative growth control, including growth arrest and apoptosis, must be understood to comprehend how appropriate cell numbers are maintained in normal intestinal mucosa. Alterations in any part of the equation contribute to mucosal atrophy or loss of epithelial integrity, and this has attracted considerable interest recently. It has been demonstrated for many years that inhibition of intestinal mucosal growth following polyamine depletion is not due to a simple decrease in expression of growth-promoting genes because polyamine-deficient cells continuously maintain high basal levels of c-Myc and c-Jun [315,323]. In other words, inhibition of intestinal mucosal growth following polyamine depletion is an active process and results primarily from the activation of expression of genes that are involved in growth arrest and apoptosis. Our previous studies and others show that several growth-inhibiting genes including p53, JunD, TGF-β, and ATF2 are activated following polyamine depletion and that polyamines modulate expression of these genes by controlling the stability of their mRNAs rather than gene transcription in intestinal epithelial cells [299–301,309].

Polyamine Depletion Stabilizes p53

The p53 gene encodes for a nuclear phosphoprotein, which was originally discovered as a cellular protein bound to the SV40 T antigen in transformed cells. p53 is present in low concentrations in normal cells and has a half-life of about 6–20 min. Expression of the p53 gene is highly regulated by the cell according to its state of growth, and its steady-state levels are significantly increased in certain growth conditions and in cells transformed by a variety of means including viral infection, chemical treatment and transfection of oncogenes [324,325].

The first evidence showing the role of the p53 gene expression in inhibition of GI mucosal growth following polyamine depletion is from our observations [326,327] and two other groups [328,329]. Using cultured IEC-6 cells that are derived from normal rat intestinal crypts, it has been shown that inhibition of polyamine synthesis by DFMO increases the p53 gene expression, which is associated with an increase in G1 phase growth arrest but not apoptosis [308]. Spermidine, given together with DFMO, completely prevented the increased expression of the p53 gene. The concentrations of p53 mRNA and protein in cells treated with DFMO plus spermidine were indistinguishable from those in cells grown in control cultures. The effect of polyamines on the p53 gene expression is specific, since polyamine depletion did not induce expression of the Rb gene in IEC-6 cells. On the other hand, increases in p53 following polyamine depletion are associated with an induction in its target genes including p21Waf1/Cip1 and p27Kip1 [315,328,329].

Consistent with our findings, Kramer et al. [328] and Ray et al. [329] have reported that exposure of human melanoma cells (MALME-3M cell) or IEC-6 cells to a polyamine analogue, *N*1, *N*11-diethylnorspermidine (DENSPM), not only decreases cellular polyamines but also increases the p53 and p21 gene expression as well. DENSPM is known to deplete polyamine pools by inhibiting biosynthetic enzymes and potently inducing the polyamine catabolic enzyme spermidine/spermine N^1-acetyltransferase. Treatment with DENSPM increases wild-type p53 (~10-fold at maximum), which is concomitant with an increase in p21 in MALME-3M cells. Another cyclin-dependent kinase inhibitor, p27, and cyclin D1 increase slightly, whereas proliferating cell nuclear antigen and p130 remain unchanged. Induction of p21 protein is paralleled by an increase in its mRNA, but induction of p53 protein is not, suggesting that cellular polyamines dictate transcriptional activation of the p21 gene and posttranscriptional regulation of the p53 gene. Consistent with the observations in IEC-6 cells, polyamine depletion by DENSPM in MALME cells also causes an increase in hypophosphorylated Rb protein.

It has been shown that p53 is an ephemeral protein [330]. In addition to transcriptional regulation, expression of the p53 gene is primarily regulated at the posttranscriptional level. Polyamine depletion increases the stability of p53 mRNA as measured by the mRNA half-life, but has no effect on the p53 gene transcription in IEC-6 cells [331]. p53 mRNA levels in control cells declined rapidly after inhibition of gene transcription by addition of actinomycin D. The half-life of p53 mRNA in control cells was ~45 min. However, the stability of p53 mRNA was dramatically increased by polyamine depletion with a half-life of >18 h. Increased half-life of p53 mRNA was prevented when spermidine was given together with DFMO. The half-life of p53 mRNA in cells treated with DFMO plus spermidine was ~48 min, similar to that of controls (without DFMO). In contrast, inhibition of polyamine synthesis did not increase the p53 gene transcription as measured by using nuclear run-on transcription assays. There were no significant differences in the rate of p53 gene transcription between control cells and cells exposed to DFMO in the presence or absence of spermidine for 6 days. These findings clearly indicate that polyamines regulate the p53 gene expression posttranscriptionally in intestinal epithelial cells and that depletion of cellular polyamines induces p53 mRNA levels primarily through the increase in its stability.

Our studies have also shown that polyamine depletion stabilizes p53 protein through the interaction with nucleophosmin (NPM) [332]. The levels of p53 protein in control cells declined rapidly after inhibition of protein synthesis by administration of cycloheximide. The half-life of p53 protein in control IEC-6 cells was ~15 min and increased to ~38 min in cells exposed to DFMO for 6 days. When DFMO was given together with spermidine, p53 protein levels decreased at the rate similar to that observed in controls, with a half-life of ~18 min. These data indicate that cellular polyamines are essential for the degradation of p53 protein and that induced accumulation of p53 protein in polyamine-deficient cells results, at least partially, from its protein stabilization. NPM is a multifunctional protein that was originally identified as a nucleolar protein involved in ribosome biogenesis and has been recently shown to regulate p53 activity [333]. We have demonstrated that polyamines modulate NPM activity in IEC-6 cells [332]. Depletion of cellular polyamines by DFMO stimulates expression of the NPM gene and induces nuclear translocation of NPM protein. Polyamine depletion stimulates NPM expression primarily by increasing both NPM gene transcription and its mRNA stability and induced NPM nuclear translocation through the activation of phosphorylation of mitogen-activated protein kinase kinase. Increased NPM physically interacts with p53 and forms a NPM/p53 complex in polyamine-deficient cells. Inhibition of NPM expression by small interfering RNA (siRNA) targeting of a specific site on the NPM mRNA not only destabilizes p53 as indicated by a decrease in its protein half-life but also prevents the increased p53-dependent transactivation as indicated by a decrease in p21-promoter activity.

Polyamines Modulate JunD mRNA Stability

JunD is a member of the jun family protooncogenes which are primary components of the activator protein-1 (AP-1) transcription factors [334]. The Jun proteins (c-Jun, JunB, and JunD) are basic-leucine transcription factors that can form either AP-1 homodimers (Jun/Jun) or AP-1 heterodimers with members of related Fos family (c-Fos, FosB, Fra-1 and Fra-2) or the ATF family [334]. Jun/Jun and Jun/Fos dimers bind to the TPA responsive element (TRE) TGACTCA present in many gene promoters, whereas Jun/ATF dimers bind preferentially to the cAMP responsive element (CRE) TGAC GTCA. There is increasing evidence that individual AP-1 dimers play distinct functions in different cellular contexts, including cell proliferation, growth arrest, differentiation, and apoptosis [334,335]. For example, c-jun and JunB function as immediate-early response genes and activation of these two genes enhances the transition from a quiescent state to proliferating state, indicating that c-Jun and JunB are positive AP-1 factors for cell proliferation. In contrast, the activation of JunD gene expression slows cell proliferation in some cell types and increases the percentage of population of cells arrested in the G_0/G_1 phase of the cell cycle [335,336], suggesting that JunD is a negative AP-1 factor and down-regulates the G_1 to S phase transition.

We have demonstrated that depletion of cellular polyamines is associated with an increase in JunD/AP-1 activity in intestinal epithelial cells [337]. Exposure of IEC-6 cells or Caco-2 (a human colon carcinoma cell line) to DFMO for 4 and 6 days increases AP-1 binding activity as measured by electrophoretic mobility shift assays, which is prevented by exogenous spermidine given together with DFMO. The anti-JunD antibody, when added to the binding reaction mixture, dramatically supershifts the AP-1 complexes present in IEC-6 cells exposed to DFMO for 4 and 6 days. The AP-1 activity attributed to JunD in the DFMO-treated cells is approximately one third of the total AP-1 binding activity on day 4, and about half on day 6, respectively. In control cells and cells exposed to DFMO and spermidine, the AP-1 binding activity is slightly supershifted by the anti-JunD antibody. On the other hand, addition of antibodies against c-Jun and JunB to the binding reaction mixture has no effect on the AP-1 binding activity in all 3 treatment groups. Although the anti-Fos antibody also partially supershifts the AP-1 complexes, there are no significant differences in the AP-1 activity attributed to Fos between control cells and polyamine-deficient cells. The increased AP-1 binding activities in polyamine-deficient cells are not supershifted by the anti-Myc antibody. These results indicate that the increase in AP-1 activity in polyamine-deficient cells is primarily contributed by an increase in JunD/AP-1 while c-Jun/AP-1 and JunB/AP-1 activity remains essentially decreased or unchanged.

Increased JunD/AP-1 activity following polyamine depletion is primarily due to the activation of JunD gene expression. Furthermore, polyamine depletion fails to induce JunD gene transcription but it stabilizes JunD mRNA. There are significant increases in JunD mRNA and protein in polyamine-deficient IEC-6 cells [337], although expression of the c-fos, c-jun, and JunB genes is decreased [316]. Increased JunD mRNA levels are ~6.6 times the control levels on day 4 and ~9 times on day 6 after treatment with DFMO. Increased levels of JunD mRNA are paralleled by a significant increase in JunD protein, which is clearly located in the nucleus. These increases in both JunD mRNA and protein in DFMO-treated cells are prevented by addition of exogenous spermidine. To test the possibility that the increase in JunD mRNA level in polyamine-deficient cells results from an increase in the mRNA synthesis, we have examined changes in the rate of JunD gene transcription by using nuclear run-on assay and showed that inhibition of polyamine synthesis did not increase JunD gene transcription. We also examined the rapid effect of addition of exogenous spermidine on JunD gene transcription in control cells and demonstrated that exposure of normal IEC-6 cells (without DFMO) to 5 μM spermidine for 2 and 4 h did not alter the rate of JunD mRNA synthesis. On the other hand, rates of c-Myc and c-Jun gene transcription were significantly decreased following polyamine depletion [315,338]. These results clearly indicate that cellular polyamines play only a minor role in the regulation of JunD gene transcription and that the increase in steady-state levels of JunD mRNA after polyamine depletion is related to a mechanism other than the stimulation of JunD gene transcription.

To determine the involvement of posttranscriptional regulation in this process, JunD mRNA stability was examined by measurement of the mRNA half-life [339]. As shown in Figure 15, depletion of cellular polyamines increases the stability of JunD mRNA in IEC-6 cells. In control cells, the half-life of JunD mRNA was ~50 min. However, the stability of JunD mRNA was increased by polyamine depletion with a half-life of >4 h. Spermidine, when given together with DFMO, almost completely prevented the increased half-life of JunD mRNA in polyamine-deficient cells. The half-life of JunD mRNA in cells exposed to DFMO plus spermidine was ~60 min, similar to that of controls (without DFMO). These findings indicate that polyamines regulate the JunD expression posttranscriptionally and that depletion of cellular polyamines induces JunD mRNA levels primarily by increasing its stability.

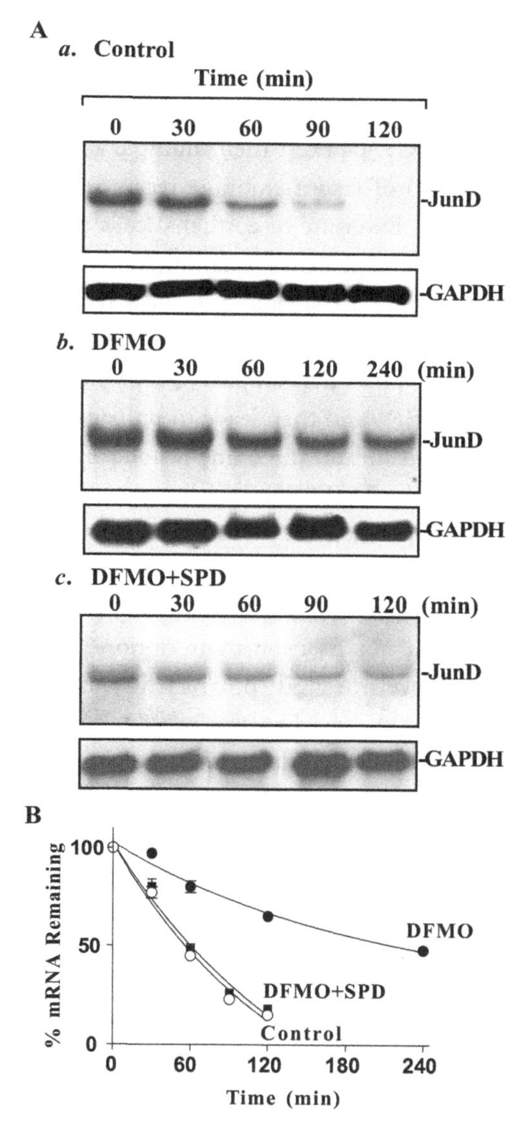

FIGURE 15: Cytoplasmic half-life studies of JunD mRNA. (A) Representative autoradiograms of Northern blots in controls and cells treated with α-difluoromethylornithine (DFMO) alone or DFMO plus spermidine (SPD). Total RNA were harvested as indicate times after administration of actinomycin D. (B) Percentage of JunD mRNA remaining after polyamine depletion. Used with permission from *Gastroenterology* 123: pp. 764–9, 2002.

Polyamine Depletion Stabilizes TGF-β mRNA and Activates Smad Signaling

TGF-β family is a group of multifunctional peptides involved in the regulation of epithelial cell growth and phenotype [340]. There are three distinct but highly related mammalian isoforms of TGF-βs, namely β1, β2, and β3. TGF-βs exert their multiple actions through heteromeric complexes of two types (types I and II) of transmembrane receptors with a serine/threonine kinase domain in their cytoplasmic region. Exposure of epithelial cells to TGF-βs leads to inhibition of growth, induction of extracellular matrix protein formation, modulation of proteolysis, and stimulation of cell migration [340]. To initiate the signaling of these responses, TGF-β binds directly to the TGF-β type II receptor (TGFβRII) that is a constitutive active kinase, after which the TGF-β type I receptor (TGFβRI) is recruited into the complex [341]. The TGFβRII in the complex phosphorylates the GS domain of TGFβRI and then leads to propagation of further downstream signals. Mutational analyses altering serine and threonine residues in the TGFβRI GS domain have indicated that the phosphorylation by TGFβRII is indispensable for TGF-β signaling, although its signaling activity does not appear to depend on the phosphorylation of any particular serine or threonine residue in the TTSGSGSG sequence of the GS domain [340,341].

We have reported that polyamine depletion activates expression of the TGF-β gene through stabilization of TGF-β mRNA but not its gene transcription [201–203]. Depletion of cellular polyamines increases the mRNA levels of TGF-β, which is paralleled by an increase in TGF-β content (Figure 16A). To determine the mechanisms by which polyamine depletion induces the TGF-β gene expression, our results show that depletion of cellular polyamines has no effect on the rate of TGF-β gene transcription, but significantly increases the half-life of mRNA for TGF-β. In control cells, the half-life of TGF-β mRNA is ~65 min, while the stability of TGF-β mRNA in polyamine-deficient cells is dramatically increased, with a half-life of >16 h. This increased stability of TGF-β mRNA in DFMO-treated cells is prevented when spermidine is given together with DFMO. These results clearly indicate that polyamines affect the TGF-β gene posttranscriptionally rather than transcriptionally, and that polyamine depletion induces the activation of TGF-β expression by increasing TGF-β mRNA stability.

Increased stabilization of TGF-β is shown to activate Smad signaling pathway in intestinal epithelial cells following polyamine depletion [201]. The Smad proteins are a family of transcriptional activators that are critical for transmitting the TGF-β superfamily signals from the cell surface to the nucleus [342]. Based on distinct functions, Smads are grouped into three classes: the receptor-regulated Smads (R-Smads), Smad2 and Smad3; the common-Smad (co-Smad), Smad4; and the inhibitory Smads (I-Smads), Smad6 and Smad7. All TGF-β family members, including TGF-βs, activins and bone morphogenetic proteins, use TGFβRI and TGFβRII receptors in a variety of cell types. Upon ligand binding, the activated TGFβRII kinase phosphorylates the TGFβRI receptor, which subsequently phosphorylates the R-Smads on a C-terminal SSXS motif.

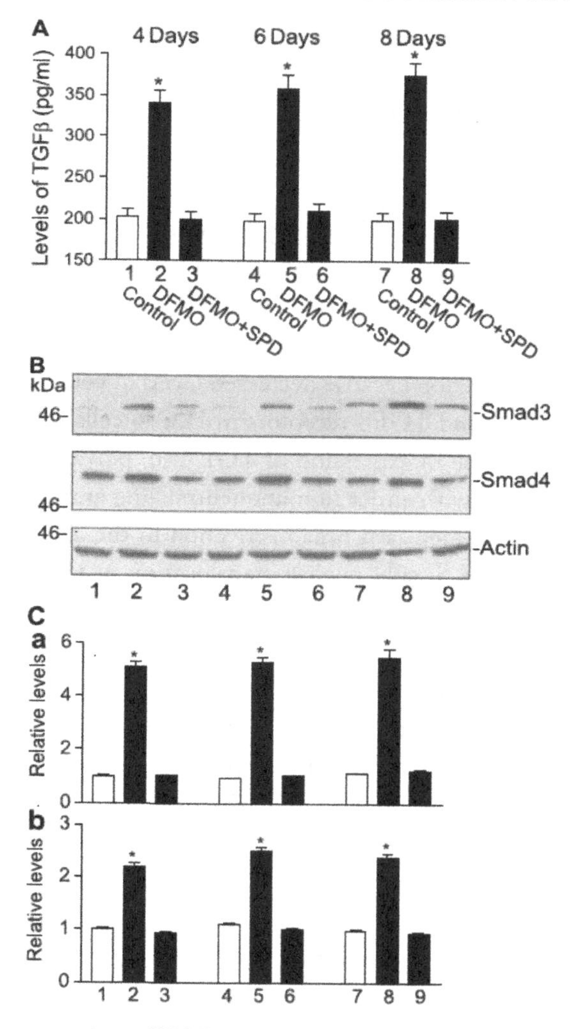

FIGURE 16: Changes in expression of TGF-β and Smad3/Smad4 after polyamine depletion. (A) Levels of TGF-β in cells treated with either α-difluoromethylornithine (DFMO) alone or DFMO plus spermidine (SPD) for various times. (B) Immunoblots of Smnad3/Smad4 proteins. (C) Quantitative analysis of Smad3/Smad4 protein levels. Used with permission from *Am J Physiol* 285: pp. G1056–67, 2003.

This induces dissociation of the R-Smad from the receptor, stimulates the assembly of a heteromeric complex between the phosphorylated R-Smads and Smad4, and results in the nuclear accumulation of this heteromeric Smad3/Smad4 complex [342,343]. In the nucleus, Smads bind to a specific DNA site (GTCTAGAC) and cooperate with various transcription factors in regulating target gene expression.

We have found that depletion of cellular polyamines by exposure to DFMO increased levels of both Smad3 and Smad4 proteins (Figure 16B and C) and induced their nuclear translocation [201]. Smad3 is shown to be highly expressed in intestinal epithelial cells and activation of this R-Smad is ligand-specific. It is not surprising that polyamine depletion increases Smad3 protein and enhances its nuclear translocation because decreased levels of cellular polyamines are known to stimulate expression of TGF-β and TGFβ receptors in IEC-6 cells [201–203]. Activated Smad3 results primarily from the increase in expression of TGF-β in polyamine-deficient cells, since inhibition of TGF-β by treatment with either immunoneutralizing anti-TGF-β antibody or TGF-β antisense oligomers prevents the increased Smad activation in the absence of cellular polyamines. Polyamine depletion also induces Smad4 nuclear translocation in IEC-6 cells [201] (Figure 16). Smad4 functions as a common mediator for all R-Smads and forms heteromeric complexes with Smad3 after ligand activation [201]. The observed change in Smad4 in polyamine-deficient cells, however, seems to be a secondary response to the activation of Smad3. In support of this possibility, treatment with exogenous TGF-β did not alter levels of Smad4 protein in normal IEC-6 cells (without DFMO), although it significantly increased Smad3 expression. Furthermore, exposure to immunoneutralizing anti-TGF-β antibody or TGF-β antisense oligomers did not prevent the increased levels of Smad4 protein in polyamine-deficient cells. The other possibility also exists, that polyamine depletion induces Smad4 expression through a mechanism independent from the activated Smad3. In addition, this increased Smad expression and nuclear translocation in the DFMO-treated cells are specifically related to polyamine depletion rather than to a nonspecific effect of DFMO because the stimulatory effect of this compound on Smads was completely prevented by the addition of exogenous spermidine.

Furthermore, polyamine depletion-induced Smad activity is associated with a significant increase in the transcriptional activation of Smad-driven promoters. Using electrophoretic mobility shift method and luciferase reporter assays, we have demonstrated that polyamine depletion increases Smad sequence-specific DNA binding and induces luciferase reporter activity of Smad-dependent promoters [202]. Our studies also show that increased transcriptional activation following polyamine depletion is primarily due to the function of Smad3/Smad4 heteromeric complexes because ectopic expression of a dominant negative mutant Smad4 prevented the increased Smad transcriptional activation in polyamine-deficient cells. These findings clearly indicate that TGF-β stabilization following polyamine depletion induces the formation of Smad3/Smad4 heteromeric complexes and activates transcription of Smad target genes, contributing to the inhibition of intestinal epithelial cell proliferation.

Polyamines Regulate Apoptosis by Altering the Stability of ATF-2 and XIAP mRNAs

The activating transcription factor-2 (ATF_2) is a member of the ATF/CRE-binding protein family of transcription factors, whereas the X chromosome-linked inhibitor of apoptosis protein (XIAP) is a potent intrinsic caspase inhibitor. We have recently demonstrated that polyamines negatively regulate expression of ATF-2 and XIAP posttranscriptionally and that polyamine-modulated ATF-2 and XIAP play an important role in the regulation of intestinal epithelial cell apoptosis [300,344,345]. Inhibition of polyamine synthesis by DFMO increased expression of the ATF-2 and XIAP genes. The steady-state levels of ATF-2 and XIAP mRNAs and proteins increased significantly in cells treated with DFMO for 4 and 6 days, but this induction was completely prevented by the addition of exogenous putrescine (10 μM) given together with DFMO. Spermidine (5 μM) had an effect equal to that of putrescine on levels of ATF-2 and XIAP mRNAs and proteins when it was added to cultures that contained DFMO.

These stable ODC-IEC cells exhibited very high levels of ODC protein and a greater than 50-fold increase in ODC enzyme activity. Accordingly, the levels of putrescine, spermidine, and spermine in ODC-IEC cells were increased by ~12-fold, ~2-fold, and ~25% when compared with cells transfected with the control vector lacking ODC cDNA [301,344]. Interestingly, ODC-IEC cells displayed a substantial decrease in expression of ATF-2 and XIAP. The levels of ATF-2 and XIAP proteins were decreased by >80% in stable ODC-IEC cells as compared with those observed in cells transfected with the control vector. The effects of ODC overexpression on the expression of ATF-2 and XIAP are not simply due to clonal variation since two stable clones, ODC-IEC-C1 and ODC-IEC-C2, showed similar responses. These results indicate that increasing cellular polyamines represses expression of ATF-2 and XIAP, while decreasing the levels of cellular polyamines increases ATF-2 and XIAP levels.

Depletion of cellular polyamines by DFMO did not change ATF-2 and XIAP gene transcription as measured by ATF-2- and XIAP-promoter luciferase reporter gene assays in untreated cells compared with cells exposed to DFMO in the presence or absence of putrescine for 4 and 6 days. Analysis of the kinetics of adding exogenous putrescine or spermidine on ATF-2- or XIAP-promoter activity in control cells revealed that exposure of normal IEC-6 cells (without DFMO) to 10 μM putrescine or 5 μM spermidine for 2 and 4 h similarly failed to alter the levels of ATF-2- and XIAP-promoter luciferase reporter activities. In contrast, depletion of cellular polyamines by DFMO significantly increased the stability of ATF-2 and XIAP mRNAs. The half-lives of ATF-2 and XIAP increased significantly in polyamine-deficient cells, which were completely prevented by exogenous putrescine. The stability of ATF-2 and XIAP mRNAs in cells exposed to DFMO plus putrescine was similar to that of control cells (without DFMO). These findings indicate that polyamine depletion induces the levels of ATF-2 and XIAP mRNAs primarily through the induction in their stability.

Polyamines are shown to regulate apoptosis through multiple signaling pathways; depletion of cellular polyamines promotes the resistance to apoptosis in normal IECs. Although silencing ATF-2 or XIAP failed to directly induce apoptosis in polyamine-deficient cells, the increased resistance to TNF-α/cycloheximide (CHX)-induced apoptosis was lost when expression of ATF-2 or XIAP was silenced by transfection with the specific siRNAs. These results indicate that the elevation in levels of ATF-2 and XIAP promote an increase in resistance to apoptosis following polyamine depletion.

Recently, polyamines are also shown to regulate MEK-1 (mitogen-activated protein kinase kinase-1) expression posttranscriptionally; and depletion of cellular polyamines stabilizes MEK-1 mRNA, increases MEK-1 protein levels, and induces to the resistance of intestinal epithelial cells to apoptosis [346].

POLYAMINES MODULATE THE STABILITY OF mRNAs VIA THE RNA-BINDING PROTEIN HuR

The mRNA turnover is primarily controlled through the association of RNA-binding proteins (RBPs) that bind to specific RNA sequences and either increase or decrease transcript half-life, thus altering the profiles of expressed gene products [345,347,348]. The best characterized *cis*-acting elements of mRNA turnover are U-rich and AU-rich sequences that are usually located in the 3'-untranslated regions [3'-UTR) of many labile mRNAs [348]. Among the RBPs that regulate specific subsets of mRNAs are several RBPs that modulate mRNA turnover (HuR, NF90, AUF1, BRF1, TTP, KSRP) and RBPs that modulate translation (HuR, TIAR, NF90, TIA-1), collectively known as *t*ranslation and *t*urnover-*r*egulatory (TTR)-RBPs [349]. HuR is a pivotal posttranscriptional regulator of gene expression and binds with great affinity and specificity to U-rich and AU-rich elements (AREs) in a variety of mRNAs which typically present one or several hits of a recently identified RNA motif [350]. Upon binding to a target mRNA, HuR has been shown to stabilize the mRNA, alter its translation, or carry out both functions [348,350]. HuR is a ubiquitously expressed member of the Hu/ELAV (embryonic lethal abnormal vision in *Drosophila melanogaster*) family of RNA-binding proteins that also comprises the primarily neuronal members HuB, HuC, and HuD [347,348]. HuR is predominantly nuclear in unstimulated cells, but it rapidly translocates to the cytoplasm in response to various stimuli [351]. Although the precise processes regulating HuR function remain to be not fully understood, it is clear that its subcellular localization is intimately linked to its effects upon target transcripts.

Polyamines Modulate Subcellular Trafficking of HuR

Our previous studies [345,352] have shown that polyamine depletion enhanced the cytoplasmic accumulation of HuR, although it had no effect on the levels of total cellular HuR (Figure 17). The induction of cytoplasmic HuR occurred as early as day 4 after exposure to DFMO and remained elevated on day 6, associated with a significant decrease in nuclear HuR. Supplementation with the polyamine putrescine reversed the effects of DFMO, preventing both the accumulation of cytoplasmic HuR and the reduction in nuclear HuR. In contrast, increased cellular polyamines by ectopic expression of the ODC gene inhibit HuR cytoplasmic translocation. These results suggest that increases in cellular polyamines promote the nuclear accumulation of HuR and reduce its cytoplasmic levels in intestinal epithelial cells. We have further demonstrated that polyamines modulate subcellular distribution of HuR through AMP-activated protein kinase [353].

Induced Cytoplasmic HuR Binds to Target mRNAs in Polyamine-Deficient Cells

A series of studies from our group has demonstrated that HuR binds to mRNAs encoding p53, NPM, JunD, ATF-2, XIAP, and MEK-1 through their 3'-UTRs or/and coding regions (CRs) and that these associations increase significantly in the cytoplasm following polyamine depletion. First, we examined if these mRNAs associated with HuR by performing RNP-IP assays using anti-HuR antibody under conditions that preserved RNP integrity [346]. The interactions of these transcripts with HuR were examined by isolating RNA from the IP material and subjecting it to reverse transcription (RT), followed by either conventional PCR or real-time quantitative (q)PCR analyses. It was found that the PCR products of p53, NPM, JunD, ATF-2, XIAP, and MEK-1 were highly enriched in HuR samples compared with control IgG samples in both parental IEC-6 and differentiated IEC-Cdx2L1 cells and that these interactions were increased after polyamine depletion. In this study, the amplification of GAPDH PCR products, found in all samples as low-level contaminating housekeeping transcripts (not HuR targets), served to monitor the evenness of sample input, as reported previously [346].

Second, we used biotinylated transcripts spanning the 3'-UTRs or CRs of these transcripts in RNA pull down assays using streptavidin-coated beads and lysates prepared from either untreated or polyamine-deficient cells. Our results showed that the 3'-UTR transcripts of p53, NPM, JunD, and MEK-1 (Figure 18) readily associated with cytoplasmic HuR, as detected by Western blot analysis of the pull down products. The intensity of these bindings increased significantly when using lysates prepared from cells that were rendered polyamine-deficient by treatment with DFMO, but they were reduced when using lysates from cells treated with exogenous putrescine. On the other hand, transcripts corresponding to CRs of these mRNAs showed undetectable binding to HuR present in cytoplasmic lysates, regardless of the presence or absence of cellular polyamines. Interestingly, HuR

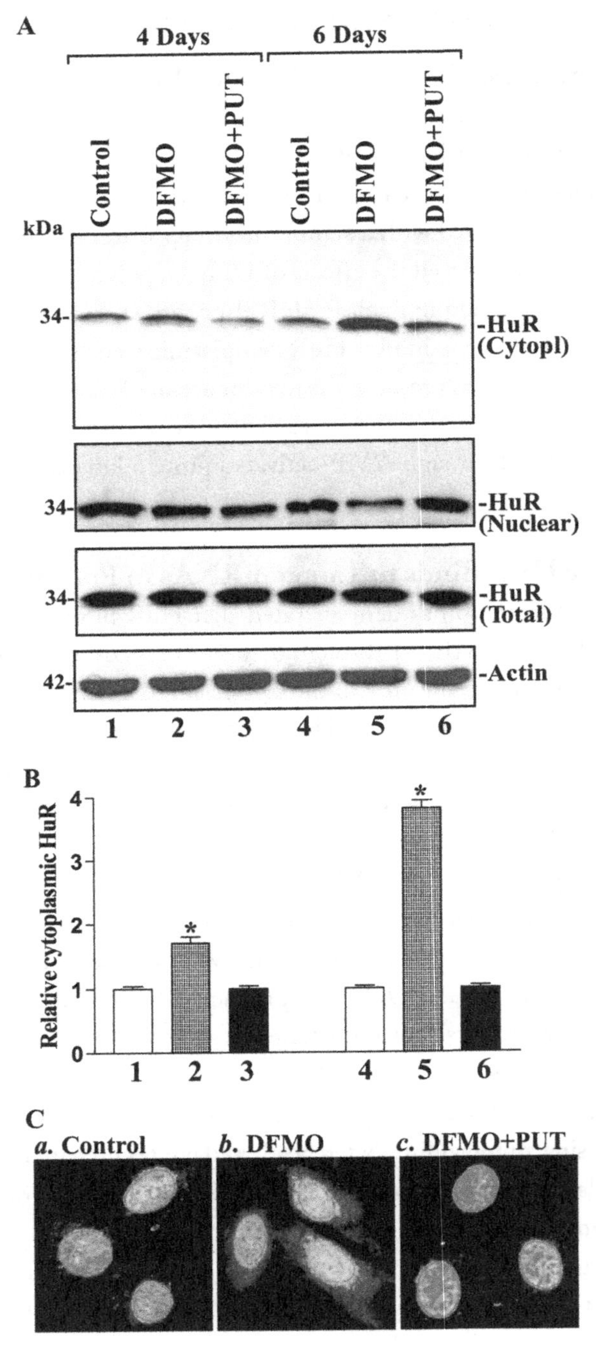

FIGURE 17: Polyamine depletion increases cytoplasmic HuR levels. (A) Levels of total HuR protein and its levels in the cytoplasm and nucleon in cells treated with either α-difluoromethylornithine (DFMO) alone or DFMO plus putrescine (PUT). (B) Quantitative analysis of HuR immunoblots. (C) HuR immunostaining. Purple, HuR; yellow, nuclei. Used with permission from *J Biol Chem* 281:19387–94, 2006.

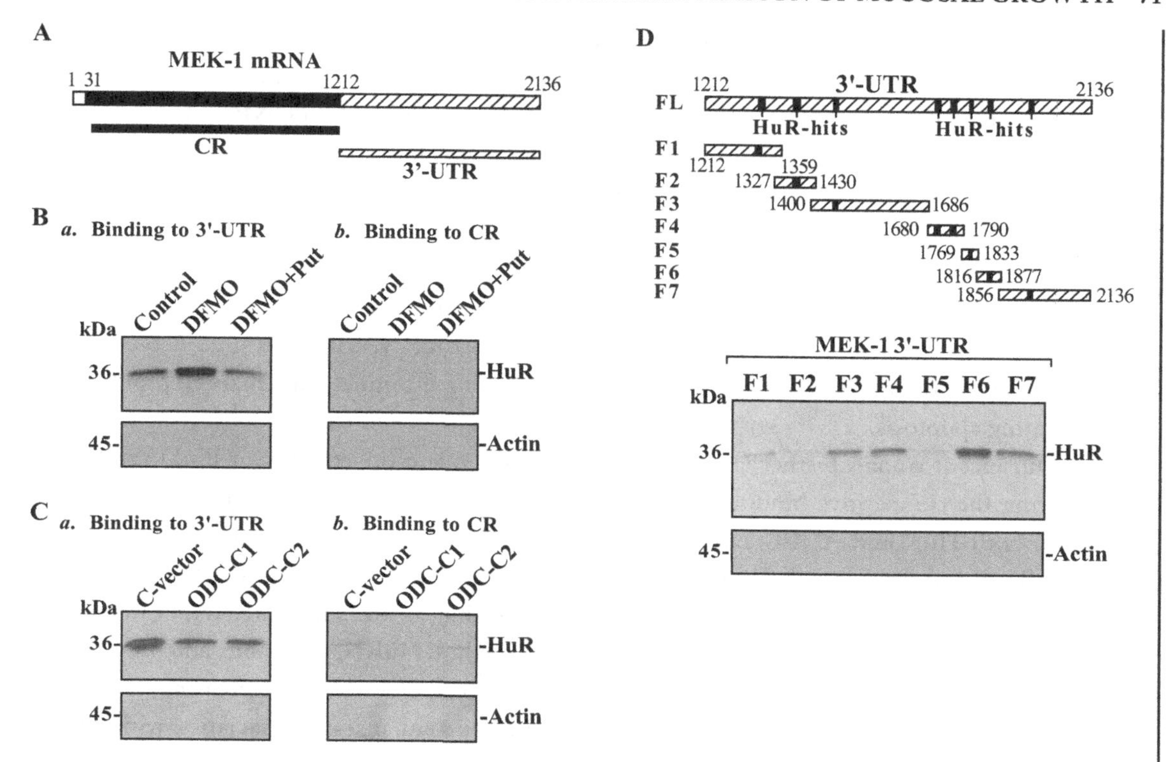

FIGURE 18: HuR-binding to the 3'-untranslated region (UTR) and coding region (CR) of MEK-1 mRNA after changing levels of cellular polyamines. (A) Schematic representation of the MEK-1 biotinylated transcripts (CR and 3'-UTR). (B) Representative HuR immunoblots using the pulldown materials by different fractions of MEK-1 mRNA after polyamine depletion. (C) Representative HuR immunoblots using the pulldown materials in clonal (C) populations of ODC-IEC cells (ODC) and control cells (C-vector). (D) Representative HuR immunoblots in the material pulled down by different biotinylated fractions of the MEK-1 mRNA 3'-UTR. Top: schematic representation of the MEK-1 3'-UTR biotinylated transcripts. Used with permission from *Biochem J* 426:293–306, 2010.

associates with both the 3'-UTR and CR of XIAP, which was increased by polyamine depletion. Together, these findings clearly show that increased cytoplasmic HuR binds to mRNAs of p53, NPM, JunD, ATF-2, XIAP, and MEK-1 following polyamine depletion.

Induced HuR Stabilizes Its Target mRNAs in Polyamine-Deficient Cells

We designed the siRNA molecule targeting the HuR mRNA (siHuR) in order to reduce HuR levels and thus directly examine its putative role in the regulation of its target mRNA stability following polyamine depletion. Transfection with siHuR potently silenced HuR expression in polyamine-deficient cells, while transfections with control siRNA (C-siRNA) showed no inhibitory effect on HuR expression. Consistently, the increase in stability of p53, NPM, JunD, ATF-2, XIAP, and MEK-1 mRNAs in polyamine-deficient cells was abolished by silencing HuR, as the half-lives of these mRNAs in DFMO-treated siRNA-transfected cells were similar to those of control cells (without DFMO). Furthermore, in HuR-silenced populations, the increased expression of proteins of p53, NPM, JunD, ATF-2, XIAP, and MEK-1 following polyamine depletion was also prevented. These findings strongly suggest that HuR critically contributes to increasing the stability of p53, NPM, JunD, ATF-2, XIAP, and MEK-1 mRNAs in polyamine-depleted cells, in turn elevating their expression and consequently inhibiting intestinal epithelial cell proliferation or regulating apoptosis.

Our recent studies further show that polyamines regulate the stability of JunD mRNA by modulating the competitive binding of its 3'-UTR to HuR and AUF1 [354]. JunD mRNA is a target of both HuR and AUF1, and depletion of cellular polyamines enhances HuR binding to JunD mRNA and decreases the levels of JunD transcript associated with AUF1, thus stabilizing JunD mRNA. The silencing of HuR increases AUF1 binding to the JunD mRNA, decreases the abundance of (HuR/JunD mRNA) complexes, renders the JunD mRNA unstable, and prevents increases in JunD mRNA and protein in polyamine-deficient cells. Conversely, increasing the cellular polyamines represses JunD mRNA interaction with HuR and enhances its association with AUF1, resulting in an inhibition of JunD expression. These results indicate that polyamines modulate the stability of JunD mRNA through HuR and AUF1 and provide new insight into the molecular functions of cellular polyamines.

mRNA TRANSLATION BY POLYAMINES

Most recently, our laboratory demonstrated that polyamines enhance HuR association with the 3'-UTR of the c-Myc mRNA by increasing HuR phosphorylation by Chk2, in turn promoting c-Myc translation [355]. Depletion of cellular polyamines inhibited Chk2 and reduced the affinity of HuR for c-Myc mRNA; these effects were completely reversed by the addition of the polyamine putrescine or by Chk2 overexpression. In cells with high content of cellular polyamines, HuR silencing or Chk2 silencing reduced c-Myc translation and c-Myc expression levels. Our findings demonstrate that polyamines regulate c-Myc translation in IECs through HuR phosphorylation by Chk2 and provide a new insight into the molecular functions of cellular polyamines in the regulation of intestinal epithelial proliferation.

In a separate study, we have shown that polyamines also regulate MEK-1 mRNA translation through HuR [346], but this effect is independent of Chk2-mediated HuR phosphorylation. Depletion of cellular polyamines enhances HuR association with MEK-1 mRNA and promotes

MEK-1 translation, whereas increasing polyamine levels represses MEK-1 translation by reducing the abundance of (HuR/MEK-1 mRNA) complex. Since HuR overexpression fails to protect against apoptosis if MEK-1 expression is silenced, these findings suggest that HuR-modulated MEK-1 translation plays a critical role in HuR-elicited anti-apoptotic programme in intestinal epithelial cells.

• • • •

Summary and Conclusions

The mammalian gut epithelium is a rapidly self-renewing tissue in the body, and its homeostasis is preserved through strict regulation of epithelial cell proliferation, growth arrest, and apoptosis. Unlike other tissues of the body, the GI mucosa is exposed to dietary constituents and its own secretions that contain a variety of molecules and factors modulating the growth of the mucosa. In addition, the food ingestion and presence of food within the digestive tract cause the release of a number of hormones, particularly gut hormones and peptide growth factors that have effects specific to the GI mucosa. Although this wide variety of growth regulators has complicated our understanding of the overall regulation of GI mucosal growth, the humoral factors have the potential to affect all cells along the GI tract equally, whereas those in the lumen lose effectiveness distally as they move down the GI tract. It is both the humoral mechanism and gradient-oriented mechanism that explain the varied growth responses to feeding, development, surgery, and disease.

An increasing body of evidence indicates that polyamines are necessary for normal intestinal mucosal growth and that decreasing cellular polyamines inhibits cell proliferation and disrupts epithelial integrity. Polyamines are shown to regulate intestinal epithelial cell renewal by virtue of their ability to modulate expression of various genes. Increasing the levels of cellular polyamines stimulates GI mucosal growth and cell proliferation by increasing expression of growth-promoting genes through enhancement of their gene transcription and translation, whereas growth inhibition following polyamine depletion results primarily from the activation of growth-inhibiting genes by stabilizing their mRNAs via the RBP HuR. In addition, polyamines also modulate apoptosis of intestinal epithelial cells through multiple signaling pathways including alterations in expression of apoptosis-associated signaling proteins.

However, there are still many critical issues that remain to be addressed regarding the roles of polyamines in maintenance of gut epithelial integrity. For example, studies to define the molecular process responsible for regulation of RNA-binding proteins by polyamines and how polyamine depletion-induced growth-inhibiting proteins interact with their downstream target signals are needed and will lead to a better understanding of the biological functions of cellular polyamines and the mechanism of polyamine depletion-induced growth arrest under physiological and various pathological conditions. It is very interesting to determine if polyamines are involved in the regulation of microRNA biogenesis in the GI tissues and how polyamine-modulated microRNAs affect expression of growth-associated genes during GI mucosal growth inhibition under various pathological conditions.

Acknowledgments

The authors sincerely apologize to all colleagues whose work has been omitted due to space limitations. Authors' work was supported by Merit Review Grants from the Department of Veterans Affairs (JNR and J-YW) and by National Institutes of Health Grants DK-57819, DK-61972, DK-68491 (J-YW). J-Y Wang is a Research Career Scientist, Medical Research Service, U.S. Department of Veterans Affairs. The authors have no conflicting financial interest.

References

[1] Silberg DG and Wu GD. Development of the alimentary tract, liver and pancreas. In: *Gastrointestinal Cancers*, edited by Rustgi AK; Saunders, 2003, Chapter 7, pp. 105–119.

[2] Antonioli DA and Madara JL. Functional anatomy of the gastrointestinal tract. In: *Pathology of Gastrointestinal Tract*, edited by Ming SC and Goldman H; Williams & Wilkins, 1998, Chapter 2, pp. 13–33.

[3] Flock MH. Stomach and duodenum. In: *Netter's Gastroenterology*, edited by Flock MH; Icon Learning Systems, 2005, Chapter 2, pp. 106–15.

[4] Klein RM. Small intestinal cell proliferation during development. In: *Human Gastrointestinal Development*, edited by Lebenthal E; Raven Press, New York, 1989, Chapter 18, pp. 367–92.

[5] Christensen J. Gross and microscopic anatomy of the large intestine. In: *The Large Intestine: Physiology, Pathophysiology, and Disease*, edited by Philips SF, Pemberton JH and Shorter RG; Raven Press, New York, 1991, Chapter 2, pp. 13–35.

[6] Montgomery RK, Mulberg AE and Grand RJ. Development of the human gastrointestinal tract: twenty years of progress. *Gastroenterology* 116: pp. 702–31, 1999.

[7] Fawcett DW. Bloom & Fawcett—a text book of histology. 11th ed., WB Saunders, Philadelphia, pp. 619–40, 1986.

[8] Gannon B. The vasculature and lymphatic drainage. In: *Gastrointestinal and Oesophageal Pathology*. 2nd ed., edited by Whitehead R; Churchill Livingstone, Edinburgh, pp. 129–99, 1995.

[9] Hitchcock RJ, Pemble MJ, Bishop AE, Spitz L, Polak JM. Quantitative study of the development and maturation of human oesophageal innervation. *J Anat* 180: pp. 175–83, 1992.

[10] Hauck AL, Swanson KS, Kenis PJA, Leckband DE, Gaskins HR, Schook LB. Twists and turns in the development and maintenance of the mammalian small intestine epithelium. *Birth Defects Res* 75: pp. 58–71, 2005.

[11] Young B, Heath J. *Wheater's Functional Histology*. 4th ed., Churchill Livingstone, Edinburgh, 2000, p. 413.

[12] Menard D. Functional development of the human gastrointestinal tract: hormone- and growth factor-mediated regulatory mechanisms. *Can J Gastroenterol* 18: pp. 39–44, 2004.

[13] Drozdowski LA, Clandinin T, Thomson ABR. Ontogeny, growth and development of the small intestine: understanding pediatric gastroenterology. *W J Gastroenterology* 16: pp. 787–99, 2010.

[14] Menard D, Arsenault P. Maturation of human fetal esophagus maintained in organ culture. *Anat Rec* 217: pp. 348–54, 1987.

[15] Geboes K, Mebis J, Desmer V. The esophagus: normal ultrastructure and pathological patterns. In: *Ultra Structure of the Digestive Tract*, edited by Motta PM and Fujita H; Martinus Nijhoff, Boston, pp. 17–34, 1988.

[16] Flock MH. Lymphatic drainage of the stomach. In: *Netter's Gastroenterology*, edited by Flock MH, Floch NR, et al.; Icon Learning Systems, Chapter 35, pp. 121–3, 2005.

[17] Aitchison M, Brown IL. Intrinsic factor in the human fetal stomach. An immunocytochemical study. *J Anat* 160: pp. 211–7, 1988.

[18] Sarles J, Moreau H, Verger R. Human gastric lipase: ontogeny and variations in children. *Acta Paediatr* 81: pp. 511–3, 1992.

[19] Zabielski R, Godlewski MM, Guilloteau P. Control of development of gastrointestinal system in neonates. *J Physiol Pharmacol* 59: pp. 35–54, 2008.

[20] Potten CS, Loeffler M. Stem cells: attributes, cycles, spirals, pitfalls and uncertainties. Lessons for and from the crypt. *Development* 110: pp. 1001–20, 1990.

[21] Potten CS. Stem cells in gastrointestinal epithelium: numbers, characteristics and death. *Philos Trans R Soc London B Biol Sci* 353: pp. 821–30, 1998.

[22] Lacroix B, Kedinger M, Simon-Assmann P, Rousset M, Zweibaum A, Haffen K. Developmental pattern of brush border enzymes in the human fetal colon. Correlation with some morphogenetic events. *Early Hum Dev* 9: pp. 95–103, 1984.

[23] Wong WM, Wright NA. Cell proliferation in gastrointestinal mucosa. *J Clin Pathol* 52: pp. 321–33, 1999.

[24] Jankowski JA, Goodlad RA, Wright NA. Maintenance of normal intestinal mucosa: function, structure, and adaptation. *Gut* Suppl 1: pp. S1–4, 1994.

[25] Stappenbeck TS, Wong MH, Saam JR, Mysorekar IU, Gordon JI. Notes from some crypt watches: regulation of renewal in the mouse intestinal epithelium. *Curr Opin Cell Biol* 10: pp. 702–9, 1998.

[26] Majumdar APN. Regulation of gastrointestinal mucosal growth during aging. *J Physiol Pharmacol* 54: pp. 143–54, 2003.

[27] Johnson LR. Regulation of gastrointestinal mucosal growth. *Physiol Rev* 68: pp. 456–502, 1988.

[28] Majumdar APN. Growth and maturation of the gastric mucosa. In: *Growth of the Intestinal Tract*, edited by Morisset J and Solomon TE; CRC Press, Chapter 7, pp. 119–30, 1991.

[29] Willems G, Galland P, Vansteenkiste Y. Cell population kinetics of zymogen and parietal cells in the stomachs of mice. *Z Zellforsch Mikrosk Anat* 134: pp. 505–18, 1972.

[30] Willems G, Galand P, Chretien J. Autoradiographic studies on cell population kinetics in dog gastric and rectal mucosa. A comparison between *in vitro* and *in vivo* methods. *Lab Invest* 23: pp. 635–9, 1970.

[31] Lawson HH. The origin of chief and parietal cells in regenerating gastric mucosa. *Br J Surg* 57: pp. 139–41, 1970.

[32] Ragins H, Wincze F, Liu SM. The origin of gastric parietal cells in the mouse. *Anat Rec* 162: pp. 99–110, 1968.

[33] Lichtenberger LM, Welsh JD, Johnson LR. Relationship between the changes in gastrin levels and intestinal properties in the starved rat. *Am J Dig Dis* 21: pp. 33–8, 1975.

[34] Johnson LR, Aures D, Yuen L. Pentagastrin-induced stimulation of protein synthesis in the gastrointestinal tract. *Am J Physiol* 217: pp. 251–4, 1969.

[35] Lehy T, Willems G. Population kinetics of antral gastrin cells in the mouse. *Gastroenterology* 71: pp. 614–9, 1976.

[36] Rao JN, Li J, Li L, Bass BL, Wang JY. Differentiated intestinal epithelial cells exhibit increased migration through polyamines and myosin II. *Am J Physiol* 277: pp. G1149–58, 1999.

[37] Suh E, Traber PG. An intestine-specific homeobox gene regulates proliferation and differentiation. *Mol Cell Biol* 16: pp. 619–25, 1996.

[38] Barker N, Clevers H. Tracking down the stem cells of the intestine: strategies to identify adult stem cells. *Gastroenterology* 133: pp. 1755–60, 2007.

[39] Weissman IL. Stem cells: units of development, units of regeneration, and units in evolution. *Cell* 100: pp. 157–68, 2000.

[40] Morshead CM, Reynolds BA, Craig CG, et al. Neural stem cells in the adult mammalian forebrain: a relatively quiescent subpopulation of subependymal cells. *Neuron* 13: pp. 1071–82, 1994.

[41] Fuchs E. The tortoise and the hair: slow-cycling cells in the stem cell race. *Cell* 137: pp. 811–9, 2009.

[42] Todaro M, Francipane MG, Medema JP, Stassi G. Colon cancer stem cells: promise of targeted therapy. *Gastroenterology* 138: pp. 2151–62, 2010.

[43] Van der Flier LG, Clevers H. Stem cells, self-renewal and differentiation in the intestinal epithelium. *Annu Rev Physiol* 71: pp. 241–60, 2009.

[44] Marshman E, Booth C, Potten CS. The intestinal epithelial stem cell. *Bioessays* 24: pp. 91–8, 2002.

[45] Barker N, Van de Wetering M, Clevers H. The intestinal stem cell. *Genes Dev* 22: pp. 1856–64, 2008.

[46] Scoville DH, Sato T, He XC, Li L. Current view: intestinal stem cells and signaling. *Gastroenterology* 134: pp. 849–64, 2008.

[47] Giannakis M, Stappenbeck TS, Mills JC, Leip DG, Lovett M, Clifton SW, Ippolito JE, Glassock JI, Armugan M, Brent MR, Gordon JI. Molecular properties of adult mouse gastric and intestinal epithelial progenitors in their niches. *J Biol Chem* 281: pp. 11292–300, 2006.

[48] Moore KA, Lemischka IR. Stem cells and their niches. *Science* 311: pp. 1880–5, 2006.

[49] Potten CS, Kovacs L, Hamilton E. Continuous labeling studies on mouse skin and intestine. *Cell Tissue Kinet* 7: pp. 271–83, 1974.

[50] Potten CS, Owen G, Booth D. Intestinal stem cells protect their genome by selective segregation of template DNA strands. *J Cell Sci* 115: pp. 2381–8, 2002.

[51] Barker N, Clevers H. Leucine-rich repeat-containing G-protein-coupled receptors as markers of adult stem cells. *Gastroenterology* 138: pp. 1681–96, 2010.

[52] Sangiorgi E, Capecchi MR. Bmi is expressed in vivo in intestinal stem cells. *Nat Genet* 40: pp. 915–20, 2008.

[53] Cheng H, Leblond CP. Origin, differentiation and renewal of the four main epithelial cell types in the mouse small intestine. Unitarian theory of the origin of the four epithelial cell types. *Am J Anat* 141: pp. 537–61, 1974.

[54] Barker N, Van Es JH, Kuipers J, Kujala P, van den Born M, Cozijnsen M, Haegebarth A, Korving J, Begthel H, Peters PJ, Clevers H. Identification of stem cells in small intestine and colon by marker gene Lgr5. *Nature* 449: pp. 1003–7, 2007.

[55] Barker N, Ridgway RA, Van Es JH, Van de Wetering M, Begthel H, van den Born M, Danenberg E, Clarke AR, Sansom OJ, Clevers H. Crypt stem cells as the cells-of-origin of intestinal cancer. *Nature* 457: pp. 608–11, 2009.

[56] Sancho E, Batlle E, Clevers H. Live and let die in the intestinal epithelium. *Curr Opin Cell Biol* 15: pp. 763–70, 2003.

[57] Wang P, Hous SX. Regulation of intestinal stem cells in mammals and *Drosophila*. *J Cell Physiol* 222: pp. 33–7, 2010.

[58] Beebe K, Lee WC, Micchelli CA. JAK/STAT signaling coordinates stem cell proliferation and multilineage differentiation in the *Drosophila* intestinal stem cell lineage. *Dev Biol* 338: pp. 28–37, 2010.

[59] Sureban SM, May R, Ramalingam S, Subramaniam D, Natarajan G, Anant S, Houchen CW. Selective blockade of DCAMKL-1 results in tumor growth arrest by a Let-7a MicroRNA-dependant mechanism. *Gastroenterology* 137: pp. 649–59, 2009.

[60] Monzo M, Navarro A, Bandres E, Artelis R, Moreno I, Gel B, Moreno J, Diaz T, Martinez

A, Belague O, Garcia-Foncillas J. Overlapping expression of microRNAs in human embryonic colon and colorectal cancer. *Cell Res* 18: pp. 823–33, 2008.

[61] Pinto D, Gregorieff A, Begthel H, Clevers H. Canonical Wnt signals are essential for homeostasis of the intestinal epithelium. *Genes Dev* 17: pp. 1709–13, 2003.

[62] Zorn AA. Wnt signaling: antagonistic Dickkopfs. *Curr Biol* 11: pp. R592–5, 2001.

[63] Lin G, Xu N, Xi R. Paracrine Wingless signaling controls self-renewal of *Drosophila* intestinal stem cells. *Nature* 455: pp. 1119–23, 2008.

[64] Batle E, Henderson JT, Begthel H, Van de Wetering M, Sancho E, Huls G, et al. β-catenin and TCF mediate cell positioning in the intestinal epithelium by controlling the expression of EphB/ephrin B. *Cell* 111: pp. 251–63, 2002.

[65] Johan H, Van Es, Philippe J, Gregorieff A, Van Gijn ME, Jonkheer S, Hatzis P, Thiele A, Van den Born M, Begthel H, Brabletz T, Taketo M, Clevers H. Wnt signaling induces maturation of Paneth cells in intestinal crypts. *Nat Cell Biol* 7: pp. 381–6, 2005.

[66] Crosnier C, Stamataki D, Lewis J. Organizing cell renewal in the intestine: stem cells, signals and combinatorial control. *Nat Rev Genet* 7: pp. 349–59, 2006.

[67] Zhu L, Gibson P, Currle DS, Tong Y, Richardson RJ, Bayazitov IT, Poppleton H, Zakharenko S, Ellison DW, Gilbertson RJ. Prominin 1 marks intestinal stem cells that are susceptible to neoplastic transformation. *Nature* 457: pp. 603–7, 2009.

[68] Barker N, Van Es JH, Kuipers J, et al. Identification of stem cells in small intestine and colon by marker gene Lgr5. *Nature* 449: pp. 1003–7, 2007.

[69] Montgomery RK, Shivadasani RA. Prominin 1 (CD133) as an intestinal stem cell marker: promise and nuance. *Gastroenterology* 136: pp. 2051–4, 2009.

[70] Potten CS, Booth C, Tudor GL, et al. Identification of a putative intestinal stem cell and early lineage marker: musashi-1. *Differentiation* 71: pp. 28–41, 2003.

[71] Haramis AP, Begthel H, Van de Born M, et al. De novo crypt formation and juvenile polyposis on BMP inhibition in mouse intestine. *Science* 303: pp. 1684–6, 2004.

[72] Batts LE, Polk DB, Dubois RN, et al. Bmp signaling is required for intestinal growth and morphogenesis. *Dev Dyn* 235: pp. 1563–70, 2006.

[73] He XC, Zhang J, Tong WG, et al. BMP signaling inhibits intestinal stem cell self-renewal through suppression of Wnt-β-catenin signaling. *Nat Genet* 36: pp. 1117–21, 2004.

[74] Howe JR, Bair JL, Sayed MG, et al. Germline mutations of the gene encoding bone morphogenetic protein receptor 1A in juvenile polyposis. *Nat Genet* 28: pp. 184–7, 2001.

[75] Ohistein B, Spradling AC. Multipotent *Drosophila* intestinal stem cells specify daughter cell fates by differential Notch signaling. *Science* 315: pp. 988–92, 2007.

[76] Artavanis-Tsakonas S, Ronald MD, Lake RJ. Notch signaling: cell fate control and signal integration in development. *Science* 284: pp. 770–6, 1999.

[77] Milano J, McKay J, Dagenais C, Foster-Brown L, Pognan F, et al. Modulation of notch

processing by γ-secretase inhibitors causes intestinal goblet cell metaplasia and induction of genes known to specify gut secretory lineage differentiation. *Toxicol Sci* 82: pp. 341–58, 2004.

[78] Fuller MT, Sradling AC. Male and female *Drosophila* germ line stem cells: two versions of immortality. *Science* 316: pp. 402–4, 2007.

[79] Gregory L, Came PJ, Brown S. Stem cell regulation by JAK/STAT signaling in *Drosophila*. *Semin Cell Dev Biol* 19: pp. 407–13, 2008.

[80] Deccota E, Spradling AC. The Drosophila ovarian and testis stem cell niches: similar somatic stem cells and signals. *Dev Cell* 9: pp. 501–10, 2005.

[81] Marinissen MJ, Gutkind JS. G-Protein coupled receptors and signaling networks: emerging paradigms. *Trends Pharmacol Sci* 22: pp. 368–76, 2001.

[82] Walsh JH. Gastrointestinal hormones. In: *Physiology of the Gastrointestinal Tract*. 3rd ed., edited by Johnson LR, Alpers DH, Christensen J, Jacobson ED, and Walsh JH; Raven Press, New York, pp. 1–128, 1994.

[83] Drucker DJ. Biological actions and therapeutic potential of the glucagon-like peptides. *Gastroenterology* 122: pp. 531–44, 2002.

[84] Thomas RP, Hellmich MR, Townsend CM, Evers BM. Role of gastrointestinal hormones in the proliferation of normal and neoplastic tissues. *Endocrine Rev* 24: pp. 571–99, 2003.

[85] Dockray GJ. Cholecystokinin and gut–brain signaling. *Regul Pept* 155: pp. 6–10, 2009.

[86] Nilsson AH. The gut as the largest endocrine organ in the body. *Ann Oncol* 12: pp. S63–8, 2001.

[87] Edkins JS. The chemical mechanism of gastric secretion. *J Physiol* 34: pp. 133–44, 1906.

[88] Gregory RA, Tracy HJ. The constitution and properties of two gastrins extracted from hog antral mucosa. *Gut* 5: pp. 105–14, 1964.

[89] Dimaline R, Varro A. Attack and defense in the gastric epithelium—a delicate balance. *Exp Physiol* 92: pp. 591–601, 2007.

[90] Dockry G, Dimaline R, Varro A. Gastrin: old hormone, new functions. *Pflugers Arch* 449: pp. 344–55, 2005.

[91] Jain RN, Samuelson LC. Differentiation of the gastric mucosa: role of gastrin in gastric epithelial cell proliferation and maturation. *Am J Physiol* 291: pp. G762–5, 2006.

[92] Brand SJ, Andersen BN, Rehfeld JF. Complete tyrosine-O-sulphation of gastrin in neonatal rat pancreas. *Nature* 309: pp. 456–8, 1984.

[93] Larsson LI, Rehfeld JF. Pituitary gastrins occur in corticotrophs and melanotrophs. *Science* 213: pp. 768–70, 1981.

[94] Brand SJ, Klarlund J, Schwartz TW, Rehfeld JF. Biosynthesis of tyrosine-O-sulfated gastrins in rat antral mucosa. *J Biol Chem* 259: pp. 13246–52, 1984.

[95] Evers BM, Townsend CM. Growth factors, hormones and receptors in GI cancers. In:

Molecular Mechanisms in Gastrointestinal Cancer, edited by Evers BM; Landes, Austin, TX, pp. 1–19, 1999.

[96] Bundgaard JR, Rehfeld JF. Posttranscriptional processing of progastrin. *Results Probl Cell Differ* 34: pp. 207–20, 2010.

[97] Evers BM. Gastrointestinal growth factors and neoplasia. *Am J Surg* 190: pp. 279–84, 2005.

[98] Walsh JH. Gastrin. In: *Gut Peptides: Biochemistry and Physiology*, edited by Walsh JH and Dockray GJ; Raven Press, New York, pp. 75–121, 1994.

[99] Dockry G, Dimaline R, Varro A. Gastrin: old hormone, new functions. *Pflugers Arch* 449: pp. 344–55, 2005.

[100] Schubert ML. Gastric exocrine and endocrine secretion. *Curr Opinion Gastroenterol* 25: pp. 529–36, 2009.

[101] Johnson LR , Aures D, Hakanson R. Effect of gastrin on the in vivo incorporation of 14C-leucine into protein of the digestive tract. *Proc Soc Exp Biol Med* 132: pp. 996–8, 1969.

[102] Crean GP, Marshall MW, Rumsey RDE. Parietal cell hyperplasia induced by the administration of pentagastrin (ICI 50, 123) to rats. *Gastroenterology* 57: pp. 147–55, 1969.

[103] Johnson LR. The trophic action of gastrointestinal hormones. *Gastroenterology* 70: pp. 278–88, 1976.

[104] Borch K, Renvall H, Liedberg G. Gastric endocrine cell hyperplasia and carcinoid tumors in pernicious anemia. *Gastroenterology* 88: pp. 638–48, 1985.

[105] Chen D, Zhao CM, Dockray GJ, Varro A, Van Hoek A, Sinclair NF, Wang TC, Koh TJ. Glycine-extended gastrin synergizes with gastrin 17 to stimulate acid secretion in gastrin-deficient mice. *Gastroenterology* 119: pp. 756–65, 2000.

[106] Hansen FL, Sundler F, Li Y, Gillespie PJ, Saunders TL, Greenson JK, Owyang C, Rehfeld JF, Samuelson LC. Impaired gastric acid secretion in gastrin deficient mice. *Am J Physiol* 274: pp. G561–8, 1998.

[107] Nagata A, Ito M, Iwata N, Kuno J, Takano H, Minowa O, Chihara K, Matsui T, Noda T. G protein-coupled cholecystokinin-β gastrin receptors are responsible for physiological cell growth of stomach mucosa in vivo. *Proc Natl Acad Sci U S A* 93: pp. 11825–30, 1996.

[108] Koh TJ, Dockray GJ, Varro A, Cahill RJ, Dangler CA, Fox JG, Wang TC. Overexpression of glycine-extended gastrin in transgenic mice results in increased colonic proliferation. *J Clin Invest* 103: pp. 1119–26, 1999.

[109] Friss-Hansen L. Lessons from the gastrin and gastrin receptor knockout mice. *Scand J. Clin Lab Invest Suppl* 234: pp. 41–6, 2001.

[110] Hoosein NM, Kiener PA, Curry RC, Rovati LC, McGilbra DK, Brattain MG. Antiproliferative effects of gastrin receptor antagonists and antibodies to gastrin on human colon carcinoma cell lines. *Cancer Res* 48: pp. 7179–83, 1988.

[111] Grabowska AM, Watson SA. Role of gastrin peptides in carcinogenesis. *Cancer Lett* 257: pp. 1–15, 2007.

[112] Ivy AC, Oldberg E. A hormone mechanism for gall bladder contraction and evacuation. *Am J Physiol* 76: pp. 599–613, 1928.

[113] Logsdon CD. Stimulation of pancreatic acinar cell by CCK, epidermal growth factor, and insulin *in vivo*. *Am J Physiol Gastrointest Liver Physiol* 251: pp. G487–94, 1986.

[114] Green GM, Levan VH, Lidde RA. Plasma cholecystokinin and pancreatic growth during adaption to dietary protein. *Am J Physiol* 251: pp. G170–4, 1986.

[115] Johnson LR, Guthrie PD. Effect of cholecystokinin and 16, 16-dimethyl prostaglandin E2 on RNA and DNA of gastric and duodenal mucosa. *Gastroenterology* 70: pp. 59–65, 1976.

[116] Hughes CA, Bates T, Dowling RH. Cholecystokinin and secretin prevent the intestinal mucosal hypoplasia of total parenteral nutrition in the dog. *Gastroenterology* 75: pp. 34–41, 1978.

[117] Fine H, Levine GM, Shiau YF. Effects of cholecystokinin and secretin on intestinal structure and function. *Am J Physiol Gastrointest Liver Physiol* 245: pp. G358–63, 1983.

[118] Stange EF, Schneider A, Seiffer E, Ditschuneit H. Effect of pentagastrin, secretin and cholecystokinin on growth and differentiation in organ cultured rabbit small intestine. *Horm Metab Res* 18: pp. 303–7, 1986.

[119] Read N, French S, Cunningham K. The role of the gut in regulating food intake in man. *Nutr Rev* 52: pp. 1–10, 1994.

[120] Berna Marc J and Jensen Robert T. Role of CCK/gastrin receptors in gastrointestinal/metabolic diseases and results of human studies using gastrin/CCK receptor agonists/antagonists in these diseases. *Curr Top Med Chem* 7: pp. 1211–31, 2007.

[121] Williams JA. Intracellular signaling mechanisms activated by cholecystokinin-regulating synthesis and secretion of digestive enzymes in pancreatic acinar cells. *Annu Rev Physiol* 63: pp. 77–97, 2001.

[122] Bragado MJ, Groblewski GE, Williams JA. p70s6k is activated by CCK in rat pancreatic acini. *Am J Physiol* 273: pp. C101–9, 1997.

[123] Bragado MJ, Groblewski GE, Williams JA. Regulation of protein synthesis by cholecystokinin in rat pancreatic acini involves PHAS-I and the p70-S6 kinase pathway. *Gastroenterology* 115: pp. 733–42, 1998.

[124] Bayliss HP, Starling EH. Mechanism of pancreatic secretion. *J Physiol Lond* 28: pp. 325–53, 1902.

[125] William CY, Ta-Min C. Secretin, 100 years later. *J Gastroenterol*, 38: pp. 1025–35, 2003.

[126] Green GM, Lyman RL. Feedback regulation of pancreatic enzyme as a mechanism for trypsin inhibitor-induced hypersecretion in rats. *Proc Soc Exp Biol Med* 140: pp. 6–12, 1972.

[127] Johnson LR, Guthrie PD. Secretin inhibition of gastrin-stimulated deoxyribonucleic acid synthesis. *Gastroenterology* 67: pp. 601–6, 1974.

[128] Johnson LR, Gutherie PD. Effect of secretin on colonic DNA synthesis. *Proc Soc Exp Biol Med* 158: pp. 521–52, 1978.

[129] Vito C, Domenico D. Somatostatin and the gastrointestinal tract. C*urr Opin Endocrinol Diabetes Obes* 17: pp. 63–8, 2010.

[130] Brazeau P, Vale W, Burgus R, et al. Hypothalamic polypeptide that inhibits the secretion of immunoreactive pituitary growth hormone. *Science* 179: pp. 77–9, 1973.

[131] Pradayrol L, Jornvall H, Mutt V, et al. N-terminally extended somatostatin: the primary structure of somatostatin-28. *FEBS Lett* 109: pp. 55–8, 1980.

[132] Olias G, Viollet C, Kusserow H, et al. Regulation and function of somatostatin receptors. *J Neurochem* 89: pp. 1057–91, 2004.

[133] Ujendra K, Grant M. Somatostatin and somatostatin receptors. *Results Probl Cell Differ* 50: pp. 137–84, 2010.

[134] Yamada Y, Post SR, Wang K, et al. Cloning and functional characterization of a family of human and mouse somatostatin receptors expressed in brain, gastrointestinal tract, and kidney. *Proc Natl Acad Sci U S A* 89: pp. 251–5, 1992.

[135] Morisset J, Genik P, Lord A, Solomon TE. Effects of chronic administration of somatostatin on rat exocrine pancreas. *Regul Pept* 4: pp. 49–58, 1982.

[136] Joeri VB, Dirk A, Luc VN, Jean-Pierre T. The role(s) of somatostatin, structurally related peptides and somatostatin receptors in the gastrointestinal tract: a review. *Regul Pept.* 156: pp. 1–8, 2009.

[137] Piqueras L, Martinez V. Role of somatostatin receptors on gastric acid secretion in wild-type and somatostatin receptor type 2 knockout mice. *Naunyn Schmiedebergs Arch Pharamcol* 370: pp. 510–20, 2004.

[138] Rivard N, Guan D, Turkelson CM, Petitclerc D, Solomon TE, Morisset J. Negative control by Sandostatin on pancreatic and duodenal growth: a possible implication of insulin-like growth factor I. *Regul Pept* 34: pp. 13–23, 1991.

[139] Lowell A, Pamela UF. From somatstatin to octreotide LAR: evolution of a somatostatin analogue. *Curr Med Res Opin* 25: pp. 2989–99, 2009.

[140] Marialuisa A, Roberto B. Somatostatin analogues in the treatment of gastroenteropancreatic neuroendocrine tumors, current aspects and perspectives. *J Exp Clin Cancer Res* 29: p. 19, 2010.

[141] Kojima M, Hosoda H, Date Y, Nakazato M, Matsuo H, Kangawa K. Ghrelin is a growth-hormone-releasing acylated peptide from stomach. *Nature* 402: pp. 656–60, 1999.

[142] Brzozowski T, Konturek PC, Konturek SJ, et al. Exogenous and endogenous ghrelin in gastroprotection against stress-induced gastric damage. *Regul Pept* 120: pp. 39–51, 2004.

[143] Schubert ML. Gastrin secretion. *Curr Opin Gastroenterol* 21: pp. 636–43, 2005.

[144] De SB, Mitselos A, Depoortere I. Motilin and ghrelin as prokinetic drug targets. *Pharmacol Ther* 123: pp. 207–23, 2009.

[145] Wren AM, Small CJ, Ward HL, Murphy KG, Dakin CL, Taheri S, et al. The novel hypothalamic peptide ghrelin stimulates food intake and growth hormone secretion. *Endocrinology* 141: pp. 4325–8, 2000.

[146] Broglio F, Gottero C, Van Koetsveld P, et al. Acetylcholine regulates ghrelin secretion in humans. *J Clin Endocrinol Metab* 89: pp. 2429–33, 2004.

[147] Lippl F, Kircher F, Erdmann J, et al. Effect of GIP, GLP-1, insulin and gastrin on ghrelin release in the isolated rat stomach. *Regul Pept* 119: pp. 93–8, 2004.

[148] Taheri S, Lin L, Austin D, Young T, Mignot E. Short sleep duration is associated with reduced leptin, elevated ghrelin and increased body mass index. *PLoS Med* 1: e62, 2004.

[149] Neary MT, Batterhaam RL. Gut hormones: implications for the treatment of obesity. *Pharmacol Ther* 124: pp. 44–56, 2009.

[150] Kotunia A, Zabielski R. Ghrelin in the postnatal development of the gastrointestinal tract. *J Physiol Pharmacol* 5: pp. 97–111, 2006.

[151] Kotunia A, Wolinski J, Slupecka M, et al. Exogenous ghrelin retards the development of the small intestine in pig neonates fed with artificial milk formula. *10th International Symposium on Digestive Physiology in Pigs. Vejle, Denmark*, 4.02: 82, 2006.

[152] Zhao Z, Sakai T. Characteristics features of ghrelin cells in the gastrointestinal tract and the regulation of stomach ghrelin expression and production. *World J Gastroenterology* 14: pp. 6306–11, 2008.

[153] Toogood A, Thornet MO. Ghrelin, not just another growth hormone secretagogue. *Clin Endocrinol* 55: pp. 589–91, 2001.

[154] Kojima M, Kangawa K. Structure and function of ghrelin. *Results Probl Cell Differ*, 46: pp. 89–115, 2008.

[155] Erspamer V, Erspamer GF, Inselvini M. Some pharmacological actions of alystesin and bombesin. *J Pharm Pharmacol* 22: pp. 875–6, 1970.

[156] Greeley GH Jr, Partin M, Spannage A, et al. Distribution of bombesin-like peptides in the alimentary canal of several vertebrate species. *Regul Pept* 16: pp. 169–81, 1986.

[157] West Sonlee D and Mercer David W. Bombesin-induced gastroprotection. *Ann Surg* 241: pp. 227–31, 2005.

[158] Erspamer V, Erspamer GF, Inselvini M, et al. Occurrence of bombesin and alytesin in extracts of the skin of three European discoglossid frogs and pharmacological actions of bombesin on extravascular smooth muscle. *Br J Pharmacol* 45: pp. 333–48, 1972.

[159] Dembinski A, Konturek PK, Konturek SJ. Role of gastrin and cholecystokinin in the

growth-promoting action of bombesin on the gastroduodenal mucosa and the pancreas. *Regul Pept* 27: pp. 343–54, 1990.

[160] Lehy T, Puccio F, Chariot J, Labeille D. Stimulating effect of bombesin on the growth of gastrointestinal tract and pancreas in suckling rats. *Gastroenterology* 90: pp. 1942–9, 1986.

[161] Chu KU, Higashide S, Ever BM, Ishizuka J, Townsend Jr CM, Thompson JC. Bombesin stimulates mucosal growth in jejunal and ileal Thiry–Vella fistulas. *Ann Surg* 221: pp. 602–11, 1995.

[162] Weber HC. Regulation and signaling of human bombesin receptors and their biological effects. *Curr Opin Endocrinol Diabetes Obes* 16: pp. 66–71, 2009.

[163] Qiao J, Kang J, Ishola TA, et al. Gastrin-releasing peptide receptor silencing suppresses and tumorigenesis and metastatic potential of neuroblastoma. *Proc Natl Acad Sci U S A* 105: pp. 12891–6, 2008.

[164] Holst JJ, Bersani M, Johnsen AH, Kofod H, Hartmann B, Orskov C. Proglucagon processing in porcine and human pancreas. *J Biol Chem* 269: pp. 18827–33, 1994.

[165] Holst JJ. Enteroglucagon. *Ann Rev Physiol.* 59: pp. 257–72, 1997.

[166] Drucker DJ, Erlich P, Asa SL, Brubaker PL. Induction of intestinal epithelial proliferation by glucagon-like peptide 2. *Proc Natl Acad Sci U S A* 93: pp. 7911–6, 1996.

[167] Litvak DA, Hellmich MR, Evers BM, Banker NA, Townsend Jr CM. Glucagon-like peptide 2 is a potent growth factor for small intestine and colon. *J Gastrointest Surg* 2: pp. 146–50, 1998.

[168] Itoh N, Obata K, Yanaihara N, Okamoto H. Human preprovasoactive intestinal polypeptide contains a novel PHI-27-like peptide, PHM-27. *Nature* 304: pp. 547–9, 1983.

[169] Johnson LR. New aspects of the trophic action of gastrointestinal hormones. *Gastroenterology* 72: pp. 788–92, 1977.

[170] Andersson S, Rosell S, Hjelmquist U, Chang D, Folkers K. Inhibition of gastric and intestinal motor activity in dogs by (Gln4) neurotensin. *Acta Physiol Scand* 100: pp. 231–5, 1977.

[171] Pederson RA. Gastric inhibitory polypeptides. In: *Gut Peptides*, edited by Walsh JH and Dockray GJ; Raven Press, New York, pp. 217–60, 1994.

[172] Samuel SM, Yung WH and Chow Billy KC. Secretin as a neuropeptide. *Mol Neurobiol*, 26: pp. 97–107, 2002.

[173] Ischia J, Patel O, Shulkes A, Baldwin GS. Gastrin-releasing peptide: different forms, different functions. *Biofactors* 35: pp. 69–75, 2009.

[174] Katoh M, Katoh M. FGF signaling network in the gastrointestinal tract (Review). *Intl J Oncol* 29: pp. 163–8, 2006.

[175] Dignass AU and Sturm A. Peptide growth factors in the intestine. *Eur J Gastroenterol Hepatol* 13: pp. 763–70, 2001.

[176] Carpenter G, Wahl MI. The epidermal growth factor family. In: *Peptides, Growth Factors and their Receptors*, edited by Sporn MB and Roberts AB; Springer-Verlag, New York, pp. 69–171, 1991.

[177] Dvork B. Milk epidermal growth factor and gut protection. *J Pediatr* 156: pp. S31–5, 2010.

[178] Carpenter G, Cohen S. Epidermal growth factor. *J Biol Chem*, 265: pp. 7709–12, 1990.

[179] Walters JRF. Cell and molecular biology of the small intestine: new insights into differentiation, growth and repair. *Curr Opin Gastroenterol* 20: pp. 70–6, 2004.

[180] Riegler M, Sedivy R, Sogukoglu T, Cosentini E, Bischof G, Teleky B, et al. Epidermal growth factor promotes rapid response to epithelial injury in rabbit duodenum *in vitro*. *Gastroenterology* 111: pp. 28–36, 1996.

[181] Dignass AU, Podolsky DK. Cytokine modulation of intestinal epithelial cell restitution: central role of transforming growth factor beta. *Gastroenterology* 105: pp. 1323–32, 1993.

[182] Berlanga-Acosta J, Playford RJ, Mandir N, Goodlad RA. Gastrointestinal cell proliferation and crypt fission are separate but complementary means of increasing tissue mass following infusion of epidermal growth factor in rats. *Gut* 48: pp. 803–7, 2001.

[183] Cellini C, Xu J, Arriaga A, Buchmiller-Crair TL. Effect of epidermal growth factor infusion on fetal rabbit intrauterine growth retardation and small intestinal development. *J Pediatr Surg* 39: pp. 891–7, 2004.

[184] Kang P, Toms D, Yin Y, Cheung Q, Gong J, De Lange K, Li J. Epidermal growth factor-expressing *Lactococcus lactis* enhances intestinal development of early-weaned pigs. *J Nutr* 140: pp. 806–11, 2010.

[185] Cheung QC, Yuan Z, Dyce PW, WU D, DeLange K, Li J. Generation of epidermal growth factor-expressing *Lactococcus lactis* and its enhancement on intestinal development and growth of early-weaned mice. *Am J Clin Nutr* 89: pp. 871–9, 2009.

[186] Berlanga-Acosta J, Gavilondo-Cowley J, Lopez-Saura P, Gonzalez-Lopez T, Castro-Santana MD, Lopez-Mola E, Guillen-Nieto G, Herrera-Martinez L. Epidermal growth factor in clinical practice—a review of its biological actions, clinical indications and safety implications. *Int Wound J* 6: pp. 331–46, 2009.

[187] Burgess AW. Growth control mechanisms in normal and transformed intestinal cells. *Phil Trans R Soc Lond* 353: pp. 903–9, 1998.

[188] Montaner B, Asbert M, Peres-Tomas R. Immunolocalization of transforming growth factor-α and epidermal growth factor receptor in the rat gastroduodenal area. *Dig Dis Sci* 44: pp. 1408–16, 1999.

[189] Schweiger M, Steffl M, Amselgruber WM. Differential expression of EGF receptor in the pig duodenum during the transition phase from maternal milk to solid food. *J Gastroenterol* 38: pp. 636–42, 2003.

[190] Kuwada SK, Li XF, Damstrup L, Dempsey PJ, Coffey RJ, Wiley HS. The dynamic expression of the epidermal growth factor receptor and epidermal growth factor ligand family in a differentiating intestinal epithelial cell line. *Growth Factors* 17: pp. 139–53, 1999.

[191] Duh G, Mouri N, Warburton D, Thomas DW. EGF regulates early embryonic mouse gut development in chemically defined organ culture. *Pediatr Res* 48: pp. 794–802, 2000.

[192] Schneider MR, Wolf E. The epidermal growth factor receptor ligands at a glance. *J Cell Physiol* 218: pp. 460–6, 2009.

[193] Taylor JA, Bernabe KQ, Guo J, Warner BW. Epidermal growth factor receptor-directed enterocyte proliferation does not induce Wnt pathway transcription. *J Pediatr Surg.* 42: pp. 981–6, 2007.

[194] Liebmann C. EGF receptor activation by GPCRs: a universal pathway reveals different versions. *Mol Cell Endocrinol* 331: pp. 222–31, 2011.

[195] Majumdar APN. Regulation of gastrointestinal mucosal growth during aging. *J Physiol Pharmacol* 54: pp. 143–54, 2003.

[196] Roberts AB, Sporn MB. Physiological actions and clinical applications of transforming growth factor-β (TGF-β). *Growth Factors* 8: pp. 1–9, 1993.

[197] Huang SS, Huang JS. TGF-β control of cell proliferation. *J Cell Biochem* 96: pp. 447–62, 2005.

[198] Eickelberg O, Centrella M, Reiss M, Kashgarian M, Wells RG. Betaglycan inhibits TGF-β signaling by preventing type-1, type-2 receptor complex formation. Glycosaminoglycan modifications alter β-glycan function. *J Biol Chem* 277: pp. 823–9, 2002.

[199] Kitisin K, Saha T, Blake T, Golestaneh N, Deng M, Kim C, Tang Y, Shetty K, Mishra B, Mishra L. TGF-β signaling in development. *Sci STKE* 399: cm 1, 2007.

[200] Lee KY, Bae S. TGF-β-dependent cell growth arrest and apoptosis. *J Biochem Mol Biol* 35: pp. 47–53, 2002.

[201] Liu L, Santora R, Rao JN, Guo X, Zou T, Zhang HM, Turner DJ, Wang JY. Activation of TGF-β-Smad signaling pathway following polyamine depletion in intestinal epithelial cells. *Am J Physiol Gastrointest Liver Physiol* 285: pp. G1056–67, 2003.

[202] Rao JN, Li L, Bass BL, Wang JY. Expression of the TGF-β receptor gene and sensitivity to growth inhibition following polyamine depletion. *Am J Physiol Cell Physiol* 279: pp. C1034–44, 2000.

[203] Patel AR, Li J, Bass BL, Wang JY. Expression of the transforming growth factor-β gene during growth inhibition following polyamine depletion. *Am J Physiol Cell Physiol* 275: pp. C590–8, 1998.

[204] Gebhardt T, Lorentz A, Detmer F, Trautwein C, Bektas H, Manns MP, Bischoff SC. Growth, phenotype and function of human intestinal mast cells are tightly regulated by transforming growth factor β1. *Gut* 54: pp. 928–34, 2005.

[205] Sangild PT, Mei J, Fowden AL, Xu RJ. The prenatal porcine intestine has low transforming growth factor-β ligand and receptor density and shows reduced trophic response to enteral diets. *Am J Physiol Regul Integr Comp Physiol* 296: pp. R1053–62, 2009.

[206] McDevitt CA, Wildey GM, Cutrone RM. Transforming growth factor-β1 in a sterilized tissue derived from the pig small intestine submucosa. *J Biomed Mater Res* 67: pp. 637–40, 2003.

[207] Hauck AL, Swanson KS, Kenis PJ, Leckband DE, Gaskins HR, Schook LB. Twists and turns in the development and maintenance of the mammalian small intestine epithelium. *Birth Defects Res C Embryo Today* 75: pp. 58–71, 2005.

[208] Thompson JS, Saxena SK, Sharp JG. Regulation of intestinal regeneration: new insight. *Microsc Res Tech* 51: pp. 129–37, 2000.

[209] Patel VB, Yu Y, Das JK, Patel BB, Majumdar AP. Schlafen-3 a novel regulator of intestinal differentiation. *Biochem Biophys Res Commun* 388: pp. 752–6, 2009.

[210] Liu X, Zhao J, Li F, Guo YS, Hellmich MR, Townsend CM Jr, Cao Y, Ko TC. Bombesin enhances TGF-β growth inhibitory effect through apoptosis induction in intestinal epithelial cells. *Regul Pept* 158: pp. 26–31, 2009.

[211] Walsh R, Seth R, Behnke J, Potten CS, Mahida YR. Epithelial stem cell-related alterations in *Trichinella spiralis*-infected small intestine. *Cell Prolif* 42: pp. 394–403, 2009.

[212] Mishra L, Shetty K, Tang Y, Stuart A, Byers SW. The role of TGF-β and Wnt signaling in gastrointestinal stem cells and cancer. *Oncogene* 24: pp. 5775–89, 2005.

[213] Podolsky DK. Regulation of intestinal epithelial proliferation: a few answers, many questions. *Am J Physiol Gastrointest Liver Physiol* 264: pp. G179–86, 1993.

[214] Lindsay CR, Evans TJ. The insulin-like growth factor system and its receptors: a potential novel anticancer target. *Biologics* 2: pp. 855–64, 2008.

[215] Humbel RE. Insulin like growth factors 1 & 2. *Eur J Biochem* 190: pp. 445–62, 1990.

[216] Singh P, Rubin N. Insulin like growth factors and binding proteins in colon cancer. *Gastroenterology*, 105: pp. 1218–37, 1993.

[217] Fruchtman S, Simmons JG, Michaylira CZ, Miller ME, Greenhalgh CJ, Ney DM, Lund PK. Suppressor of cytokine signaling-2 modulates the fibrogenic actions of GH and IGF-I in intestinal mesenchymal cells. *Am J Physiol Gastrointest Liver Physiol,* 289: pp. G342–50, 2005.

[218] Zarrilli R, Bruni CB, Riccio A. Multiple levels of control of insulin-like growth factor gene expression. *Mol Cell Endocrinol* 101: pp. R1–14, 1994.

[219] Steeb CB, Trahair JF, Read LC. Administration of insulin-like growth factor-I (IGF-I) peptides for three days stimulates proliferation of the small intestinal epithelium in rats. *Gut* 37: pp. 630–8, 1995.

[220] Ohneda K, Ulshen MH, Fuller CR, D' Ercole AJ, Lund PK. Enhanced growth of small bowel in transgenic mice expressing human insulin-like growth factor-I. *Gastroenterology* 112: pp. 444–54, 1997.

[221] Roffler B, Fah A, Sauter SN, Hammon HM, Gallmann P, Brem G, Blum JW. Intestinal morphology, epithelial cell proliferation and absorptive capacity in neonatal calves fed milk-born insulin-like growth factor-I or a colostrum extract. *J Dairy Sci,* 86: pp. 1797–806, 2003.

[222] Wheeler EE, Challacombe DN. The trophic action of growth hormone, insulin-like growth factor-I and insulin on human duodenal mucosa cultured *in vitro. Gut* 40: pp. 57–60, 1997.

[223] McDonald RS. The role of insulin-like growth factors in small intestinal cell growth and development. *Horm Metab Res* 31: pp. 103–13, 1999.

[224] Howarth GS. Insulin-like growth factor-I and the gastrointestinal system: therapeutic indications and safety implications. *J Nutr* 133: pp. 2109–12, 2003.

[225] Simmons JG, Ling Y, Wilkins H, Fuller CR, D' Ercole AJ, Fagin J, Lund PK. Cell-specific effects of insulin receptor substrate-1 deficiency on normal and IGF-I-mediated colon growth. *Am J Physiol Gastrointest Liver Physiol* 293: pp. G995–1003, 2007.

[226] Eswarakumar VP, Lax I, Schlessinger J. Cellular signaling by fibroblast growth factor receptors. *Cytokine Growth Factor Rev* 16: pp. 139–49, 2005.

[227] Dailey L, Ambrosetti D, Mansukhani A, Basilico C. Mechanisms underlying differential responses to FGF signaling. *Cytokine Growth Factor Rev* 16: pp. 233–247, 2005.

[228] Podolsky DK. Healing the epithelium: solving the problem from two sides. *J Gastroenterol* 32: pp. 122–6, 1997.

[229] Winter TA, Kidd M, Kaye P, Marks IN. Gastric and duodenal mucosal protein fractional synthesis and growth factor expression in patients with *H. pylori*-associated gastritis before and after eradication of the organism. *Dig Dis Sci* 49: pp. 925–30, 2004.

[230] Lee FS, Lane TF, Kuo A, Shackleford GM, Leder P. Insertional mutagenesis identifies a member of the Wnt gene family as a candidate oncogene in the mammary epithelium of int-2/Fgf-3 transgenic mice. *Proc Natl Acad Sci U S A* 92: pp. 2268–72, 1995.

[231] Taupin DR, Pang KC, Green SP, Giraud AS. The trefoil peptides spasmolytic polypeptide and intestinal trefoil factor are major secretory products of the rat gut. *Peptides* 16: pp. 1001–5, 1995.

[232] Milani S, Calabro A. Role of growth factors and their receptors in gastric ulcer healing. *Microsc Res Tech* 53: pp. 360–71, 2001.

[233] Naldini L, Tamagnone L, Vigna E, Sachs M, Hartmann G, Birchmeier W, Daikuhara Y, Tsubouchi H, Blasi F, Comoglio PM. Extracellular proteolytic cleavage by urokinase is

required for activation of hepatocyte growth factor/scatter factor. *EMBO J* 11: pp. 4825–33, 1992.

[234] Martins A, Han J, Kim SO. The multifaceted effects of granulocyte colony-stimulating factor in immunomodulation and potential role in intestinal immune homeostasis. *IUBMB Life* 62: pp. 611–7, 2010.

[235] Aldewachi HS, Wright NA, Appleton DR, Watson AJ. The effect of starvation and refeeding on cell population kinetics in the rat small bowel mucosa. *J Anat* 119: pp. 105–21, 1975.

[236] Jenkins AP, Thompson RPH. Enteral nutrition and the small intestine. *Gut* 35: pp. 1765–9, 1994.

[237] McCole DF, Barrett KE. Varied role of the gut epithelium in mucosal homeostasis. *Curr Opin Gastroenterol* 23: pp. 647–54, 2007.

[238] Clarke RM. 'Luminal nutrition' versus 'functional workload' as controllers of mucosal morphology and epithelial replacement in the rat small intestine. *Digestion* 15: pp. 411–24, 1977.

[239] Williamson RCN, Buchholtz TW, Malt RA. Humoral stimulation of cell proliferation in small bowel after transection and resection in rats. *Gastroenterology* 75: pp. 249–54, 1978.

[240] Bjornvad CR, Thymann T, Deutz NE, Burrin DG, Jensen SK, Jensen BB, Molbak L, Boye M, Larsson LI, Schmidt M, Michaelsen KF, Sangild PT. Enteral feeding induces diet-dependent mucosal dysfunction, bacterial proliferation and necrotizing enterocolitis in preterm pigs on parenteral nutrition. *Am J Physiol Gastrointest Liver Physiol* 295: pp. G1092–103, 2008.

[241] Shanahan F. Gut microbes: from bugs to drugs. *Am J Gastroenterol* 105: pp. 275–79, 2010.

[242] Dahly EM, Gillingham MB, Guo Z, Murali SG, Nelson DW, Holst JJ, Ney DM. Role of luminal nutrients and endogenous GLP-2 in intestinal adaption to mid-small bowel resection. *Am J Physiol Gastrointest Liver Physiol* 284: pp. G670–82, 2003.

[243] Milovic V. Polyamines in the gut lumen: bioavailability and biodistribution. *Eur J Gastroenterol Hepatol* 13: pp. 1021–5, 2001.

[244] Dowling RH. The influence of luminal nutrition on intestinal adaption after small bowel resection and by-pass. *Intestinal Adaptation*, edited by Dowling RH and Riecken EO, Schattauer, Stuttgart, FRG, pp. 35–45, 1974.

[245] Riecken EO, Menge H. Nutritive effects of food constituents on the structure and function of the intestine. *Acta Hepatogastroenterol* 24: pp. 389–99, 1977.

[246] Jacobs LR, Taylor BR, Dowling RH. Effect of luminal nutrition on the intestinal adaptation following Thiry–Vella by-pass in the dog. *Clin Sci Mol Med* 49: p. 26, 1975.

[247] Clarke RM. Luminal nutrition versus functional work load as controllers of mucosal morphology and epithelial replacement in the rat small intestine. *Digestion* 15: pp. 411–24, 1977.

[248] Philpott DJ, Kirk DR, Butzner JD. Luminal factors stimulate intestinal repair during the refeeding of malnourished infant rabbits. *Can J Physiol Pharmacol* 71: pp. 650–6, 1993.

[249] Kitchen PA, Goodlad RA, FitzGerald AJ, Mandir N, Ghatei MA, Bloom SR, Berlanga-Acosta J, Playford RJ, Forbes A, Walters JR. Intestinal growth in parenterally-fed rats induced by the combined effects of glucagon-like peptide 2 and epidermal growth factor. *J Parenter Enteral Nutr* 29: pp. 248–54, 2005.

[250] Petersen YM, Hartmann B, Holst JJ, Le Huerou-Luron I, Bjornvad CR, Sangild PT. Introduction of enteral food increases plasma GLP-2 and decreases GLP-2 receptor mRNA abundance during pig development. *J Nutr* 133: pp. 1781–6, 2003.

[251] Weser E, Babbitt J, Vandeventer A. Relationship between enteral glucose load and adaptive mucosal growth in the small bowel. *Dig Dis Sci* 30: pp. 675–81, 1985.

[252] Weser E, Babbitt J, Hoban M, Vandeventer A. Intestinal adaptation: different growth responses to disaccharides compared with monosaccharides in rat small bowel. *Gastroenterology* 91: pp. 1521–7, 1986.

[253] Alpers DH. Protein synthesis in intestinal mucosa: the effect of route administration of precursor amino acids. *J Clin Invest* 51: pp. 167–73, 1972.

[254] Hirschfield JS, Kern F. Protein starvation and the small intestine. Incorporation of orally and intraperitoneally administered 1-leucine 4, 5-3H into intestinal mucosal protein of protein deprived rats. *J Clin Invest* 48: pp. 1224–9, 1969.

[255] Maxton DG, Cynk EU, Thompson RPH. Promotion of growth of the small intestine by dietary essential fatty acids. *Eur J Gastroenterol Hepatol* 2: pp. 131–6, 1990.

[256] Vanderhoof JA, Park JHY, Mohammadpour H, Blackwood D. Effects of dietary lipids on recovery from mucosal injury. *Gastroenterology* 98: pp. 1226–31, 1990.

[257] Goodlad RA, Lenton W, Ghatei MA, Adrian TE, Bloom SR, Wright NA. Effects of an elemental diet, inert bulk and different types of dietary fiber on the response of the intestinal epithelium to refeeding in the rat and relationship to plasma gastrin, enteroglucagon, and PYY concentrations. *Gut* 28: pp. 171–80, 1987.

[258] Weser E. Nutritional aspects of malabsorption: short gut adaptation. *Clin Gastroenterol* 12: pp. 443–61, 1983.

[259] Nelson DW, Murali SG, Liu X, Koopmann MC, Holst JJ, Ney DM. Insulin like growth factor I and glucagon like peptide 2 responses to fasting followed by controlled or ad libitum refeeding in rats. *Am J Physiol Regul Integr Comp Physiol* 294: pp. R1175–84, 2008.

[260] Dahly EM, Gillingham MB, Guo Z, Murali SG, Nelson DW, Holst JJ, Ney DM. Role of luminal nutrients and endogenous GLP-2 in intestinal adaptation to mid-small bowel resection. *Am J Physiol Gastrointest Liver Physiol* 284: pp. G670–82, 2003.

[261] Awad WA, Ghareeb K, Bohm J. Effect of addition of a probiotic micro-organism to broiler

diet on intestinal mucosal architecture and electrophysiological parameters. *J Anim Physiol Anim Nutr (Berl)* 94: pp. 486–94, 2010.

[262] Abreu MT. Toll-like receptor signaling in the intestinal epithelium: how bacterial recognition shapes intestinal function. *Nat Rev Immunol* 10: pp. 131–44, 2010.

[263] Del Piano M, Morelli L, Strozzi GP, Allesina S, Barba M, Deidda F, Lorenzini P, Ballare M, Montino F, Orsello M, Sartori M, Garello E, Carmagnola S, Pagliarulo M, Capurso L. Probiotics: from research to consumer. *Dig Liver Dis* 38: Suppl 2: pp. S248–55, 2006.

[264] Ewaschuk JB, Dielman LA. Probiotics and prebiotics in chronic inflammatory bowel disease. *World J Gastroenterol* 12: pp. 5941–50, 2006.

[265] Roy CC, Kein CL, Bouthillier L, Levy E. Short-chain fatty acids: ready for prime time? *Nutr Clin Pract* 21: pp. 351–66, 2006.

[266] Lara-Villoslada F, Olivares M, Sierra S, Rodriguez JM, Boza J, Xaus J. Beneficial effects of probiotic bacteria isolated from breast milk. *Br J Nutr* 98: Suppl 1, pp. S96–100, 2007.

[267] Santosa S, Farnworth E, Jones PJ. Probiotics and their potential health claims. *Nutr Rev* 64: pp. 265–74, 2006.

[268] Neish AS. Microbes in gastrointestinal health and disease. *Gastroenterology* 136: pp. 65–80, 2009.

[269] Di Giancamille A, Vitari F, Savoini G, Bontempo V, Bersani C, Dell'Orto V, Domeneghini C. Effects of orally administered probiotic *Pediococcus acidilactici* on the small and large intestine of weaning piglets. A qualitative and quantitative micro-anatomical study. *Histol Histopathol* 23: pp. 651–64, 2008.

[270] Salminen S, Benno Y, de Vos W. Intestinal colonization, microbiota and future probiotics? *Asia Pac J Clin Nutr* 15: pp. 558–62, 2006.

[271] Awad WA, Ghareeb K, Bohm J. Effect of addition of a probiotic micro-organism to broiler diet on intestinal mucosal architecture and electrophysiological parameters. *J Anim Physiol Anim Nutr (Berl)* 94: pp. 486–94, 2010.

[272] Yan F, Cao H, Cover TL, Whitehead R, Washington MK, Polk DB. Soluble proteins produced by probiotic bacteria regulate intestinal epithelial cell survival and growth. *Gastroenterology* 132: pp. 562–75, 2007.

[273] Lin PW, Nasr TR, Berardinelli AJ, Kumar A, Neish AS. The probiotic *Lactobacillus* GG may augment intestinal host defense by regulating apoptosis and promoting cytoprotective responses in the developing murine gut. *Pediatr Res* 64: pp. 511–16, 2008.

[274] Hanaway P. Balance of flora, galt and mucosal integrity. *Altern Ther Health Med* 12: pp. 52–60, 2006.

[275] Madsen K. Probiotics and the immune response. *J Clin Gastroenterol* 40: pp. 232–4, 2006.

[276] Mathers JC. Nutrient regulation of intestinal proliferation and apoptosis. *Proc Nutr Soc* 57: pp. 219–23, 1998.

[277] Reeds PJ, Burrin DG. Glutamine metabolism: nutritional and clinical significance. glutamine and the bowel. *J Nutr* 131: pp. 2505S–8S, 2001.

[278] Schottstedt T, Muri C, Morel C, Philipona C, Hammon HM, Blum JW. Effects of feeding Vitamin A and lactoferrin on epithelium of lymphoid tissues of intestine of neonatal calves. *J Dairy Sci* 88: pp. 1050–61, 2005.

[279] Sanderson IR. Dietary regulation of genes expressed in the developing intestinal epithelium. *Am J Clin Nutr* 68: pp. 999–1005, 1998.

[280] Tannus AF, Darmaun D, Ribas DF, Oliveira JE, Marchini JS. Glutamine supplementation does not improve protein synthesis rate by the jejunal mucosa of the malnourished rat. *Nutr Res* 29: pp. 596–601, 2009.

[281] Wiren M, Magnusson KE, Larsson J. The role of glutamine, serum and energy factors in growth of enterocyte-like cell lines. *Int J Biochem Cell Biol* 30: pp. 1331–6, 1998.

[282] Tuhacek LM, Mackey AD, Li N, DeMarco VG, Stevens G, Neu J. Substitutes for glutamine in proliferation of rat intestinal epithelial cells. *Nutrition* 20: pp. 292–7, 2004.

[283] Shyntum Y, Iyer SS, Tian J, Hao L, Mannery YO, Jones DP, Ziegler TR. Dietary sulfur amino acid supplementation reduces small bowel thiol/disulphide redox state and stimulates ileal mucosal growth after massive small bowel resection in rats. *J Nutr* 139: pp. 2272–8, 2009.

[284] Fedriko V, Bostick RM, Flanders WD, Long Q, Sidelnikov E, Shaukat A, Daniel CR, Rutherford RE, Woodard JJ. Effects of vitamen D and calcium on proliferation and differentiation in normal colon mucosa: a randomized clinical trial. *Cancer Epidemiol Biomarkers Prev* 18: pp. 2933–41, 2009.

[285] Zanoni JN, Fernandes PRV. Cell proliferation of the ileum intestinal mucosa of diabetic rats treated with ascorbic acid. *Biocell* 32: pp. 163–8, 2008.

[286] Uni Z, Zaiger G, Gal-Garber O, Pines M, Rozenboim I, Reifen R. Vitamin A deficiency interferes with proliferation and maturation of cells in the chicken small intestine. *Br Poult Sci* 41: pp. 410–5, 2000.

[287] Thomas S, Prabhu R, Balasubramanian KA. Retinoid metabolism in the rat small intestine. *Br J Nutr* 93: pp. 59–63, 2005.

[288] Zanoni JN, Fernandes PRV. Cell proliferation of the ileum intestinal mucosa of diabetic rats treated with ascorbic acid. *Biocell,* 32: pp. 163–8, 2008.

[289] Mashiko T, Nagafuchi S, Kanbe M, Obara Y, Hagawa Y, Takahashi T, Katoh K. Effects of dietary uridine 5'-monophosphate on immune responses in newborn calves. *J Anim Sci* 87: pp. 1042–7, 2009.

[290] Tang Z, Yin Y, Zhang Y, Huang R, Sun Z, Li T, Chu W, Kong X, Li L, Geng M, Tu Q. Effects of dietary supplementation with an expressed fusion peptide bovine lactoferricin-lactoferrampin on performance, immune function and intestinal mucosal morphology in piglets weaned at age 21 d. *Br J Nutr* 101: pp. 998–1005, 2009.

[291] Le Gall M, Gallois M, Seve B, Louveau I, Holst JJ, Oswald IP, Lalles JP, Guilloteau P. Comparative effect of orally administered sodium butyrate before or after weaning on growth and several indices of gastrointestinal biology of piglets. *Br J Nutr* 102: pp. 1285–96, 2009.

[292] Jing MY, Sun JY, Weng XY, Wang JF. Effects of zinc levels on activities of gastrointestinal enzymes in growing rats. *J Anim Physiol Anim Nutr (Berl)* 93: pp. 606–12, 2009.

[293] Yin J, Li X, Li D, Yue T, Fang Q, Ni J, Zhou X, Wu G. Dietary supplementation with zinc oxide stimulates ghrelin secretion from the stomach of young pigs. *J Nutr Biochem* 20: pp. 783–90, 2009.

[294] Milovic V. Polyamines in the gut lumen: bioavailability and biodistribution. *Eur J Gastroenterol Hepatol* 13: pp. 1021–5, 2001.

[295] Seiler N, Raul F. Polyamines and intestinal tract. *Crit Rev Clin Lab Sci* 44: pp. 365–411, 2007.

[296] Seiler N, Raul F. Polyamines and apoptosis. *J Cell Mol Med* 9: pp. 623–42, 2005.

[297] Parveen N, Cornell KA. Methylthioadenosine/S-adenosylhomocysteine nucleosidase, a critical enzyme for bacterial metabolism. *Mol Microbiol* 79: pp. 7–20, 2011.

[298] Gawel-Thompson KJ, Greene RM. Epidermal growth factor: modulator of murine embryonic palate mesenchymal cell proliferation, polyamine biosynthesis and polyamine transport. *J Cell Physiol* 140: pp. 359–70, 1989.

[299] Wang JY. Polyamines and mRNA stability in regulation of intestinal mucosal growth. *Amino Acids* 33: pp. 241–52, 2007.

[300] Xiao L, Rao JN, Zou T, Liu L, Yu TX, Zhu XY, Donahue JM, Wang JY. Induced ATF-2 represses CDK4 transcription through dimerization with JunD inhibiting intestinal epithelial cell growth after polyamine depletion. *Am J Physiol Cell Physiol* 298: pp. C1226–34, 2010.

[301] Zhang AH, Rao JN, Zou T, Liu L, Marasa BS, Xiao L, Chen J, Turner DJ, Wang JY. P53-dependent NDRG1 expression induces inhibition of intestinal epithelial cell proliferation but not apoptosis after polyamine depletion. *Am J Physiol Cell Physiol* 293: pp. C379–89, 2007.

[302] Wang JY, McCormack SA, Viar MJ, Johnson LR. Stimulation of proximal small intestinal mucosal growth by luminal polyamines. *Am J Physiol Gastrointest Liver Physiol* 261: pp. G504–11, 1991.

[303] Wang JY, Johnson LR. Polyamines and ornithine decarboxylase during repair of duodenal mucosa after stress in rats. *Gastroenterology* 100: pp. 333–43, 1991.

[304] McCormack SA, Johnson LR. Role of polyamines in gastrointestinal mucosal growth. *Am J Physiol Gastrointest Liver Physiol* 260: pp. G795–806, 1991.

[305] Milovic V. Polyamines in the gut lumen: bioavailability and biodistribution. *Eur J Gastroenterol Hepatol* 13: pp. 1021–5, 2001.

[306] Chen J, Xiao L, Rao JN, Zou T, Liu L, Bellavance E, Gorospe M, Wang JY. JunD represses transcription and translation of the tight junction protein zona occludens-1 modulating intestinal epithelial barrier function. *Mol Biol Cell* 19: pp. 3701–12, 2008.

[307] Liu L, Rao JN, Zou T, Xiao L, Wang PY, Turner DJ, Gorospe M, Wang JY. Polyamines regulate c-Myc translation through Chk2-dependent HuR phosphorylation. *Mol Biol Cell* 20: pp. 4885–98, 2009.

[308] Li L, Li J, Rao JN, Li M-L, Bass BL, Wang JY. Inhibition of polyamine synthesis induces p53 gene expression but not apoptosis. *Am J Physiol Cell Physiol* 276: pp. C946–54, 1999.

[309] Xiao L, Rao JN, Zou T, Liu L, Marasa BS, Chen J, Turner DJ, Passaniti A, Wang JY. Induced JunD in intestinal epithelial cells represses CDK4 transcription through its proximal promoter region following polyamine depletion. *Biochem J* 403: pp. 573–81, 2007.

[310] Rao JN, Li L, Bass BL, Wang JY. Expression of the TGF-β receptor gene and sensitivity to growth inhibition following polyamine depletion. *Am J Physiol Cell Physiol* 279: pp. C1034–44, 2000.

[311] Shantz LM, Pegg AE. Translational regulation of ornithine decarboxylase and other enzymes of the polyamine pathway. *Int J Biochem Cell Biol* 31: pp. 107–22, 1999.

[312] Shantz LM, Holm I, Janne OA, Pegg AE. Regulation of S-adenoxylmethionine decarboxylase activity by alterations in the intracellular polyamine content. *Biochem J* 288: pp. 511–8, 1992.

[313] Celano P, Baylin SB, Casero RA. Polyamines differentially modulate the transcription of growth-associated genes in human colon carcinoma cells. *J Biol Chem* 264: pp. 8922–7, 1989.

[314] Wang JY, McCormack SA, Viar MJ, Wang HL, Tzen CY, Scott RE, Johnson LR. Decreased expression of protooncogenes c-fos, cmyc, and c-jun following polyamine depletion in IEC-6 cells. *Am J Physiol Gastrointest Liver Physiol* 265: pp. G331–8, 1993.

[315] Liu L, Guo X, Rao JN, Zou T, Marasa BS, Chen J, Greenspon J, Casero RA Jr, Wang JY. Polyamine-modulated c-Myc expression in normal intestinal epithelial cells regulates p21Cip1 transcription through a proximal promoter region. *Biochem J* 398: pp. 257–67, 2006.

[316] Wang JY, Johnson LR. Expression of protooncogenes *c-fos* and *c-myc* in healing of gastric mucosal stress ulcers. *Am J Physiol Gastrointest Liver Physiol* 266: pp. 878–86, 1994.

[317] Quaroni A, Wands J, Trelstad RL, Isselbacher KJ. Epithelioid cell cultures from rat small intestine: characterization by morphologic and immunologic criteria. *J Cell Biol* 80: pp. 248–65, 1979.

[318] Patel AR, Wang JY. Polyamines modulate transcription but not posttranscription of c-myc and c-jun in IEC-6 cells. *Am J Physiol Cell Physiol* 273: C1020–C1029, 1997.

[319] Hobbs CA, Gilmour SK. Role of polyamines in the regulation of chromatin acetylation. In: *Text book of Polyamine Cell Signaling*, edited by Wang JY and Casero RA Jr. Humana Press, Totowa, NJ, pp. 75–89, 2006.

[320] Gerner EW, Meyskens FL. Polyamines and cancer: old molecules, new understanding. *Nat Rev* 4: pp. 781–92, 2004.

[321] Tabor CW, Tabor TH. Polyamines. *Annu Rev Biochem* 53: pp. 749–90, 1984.

[322] Panagiotidis CA, Artandi S, Calame K, Silverstein SJ. Polyamines alter sequence-specific DNA–protein interactions. *Nucleic Acids Res* 23: pp. 1800–9, 1995.

[323] Liu L, Li L, Rao JN, Zou T, Zhang HM, Boneva D, Bernard MS, Wang JY. Polyamine-modulated expression of c-myc plays a critical role in stimulation of normal intestinal epithelial cell proliferation. *Am J Physiol Cell Physiol* 288: pp. C89–99, 2005.

[324] Deppert W. The yin and yang of p53 in cellular proliferation. *Semin Cancer Biol* 5: pp. 187–202, 1994.

[325] Harris CC. The p53 tumor suppressor gene: at the crossroads of molecular carcinogenesis, molecular epidemiology and cancer risk assessment. *Science* 262: pp. 1980–1, 1993.

[326] Li L, Li J, Rao JN, Li M, Bass BL, Wang JY. Inhibition of polyamine synthesis induces p53 gene expression but not apoptosis. *Am J Physiol Cell Physiol* 276: pp. C946–54, 1999.

[327] Li L, Rao JN, Li J, Patel AR, Bass BL, Wang JY. Polyamine depletion stabilizes p53 resulting in inhibition of normal intestinal epithelial cell proliferation. *Am J Physiol Cell Physiol* 281: pp. C941–54, 2001.

[328] Kramer DL, Vujcic S, Diegelman P, Alderfer J, Miller JT, Black JD, Bergeron RJ, Porter CW. Polyamine analogue induction of the p53-p21/WAF1/CIP1-Rb pathway and G1 arrest in human melanoma cells. *Cancer Res* 59: pp. 1278–86, 1999.

[329] Ray RM, Zimmerman BJ, McCormack SA, Patel TB, Johnson LR. Polyamine depletion arrests cell cycle and induces inhibitors p21Waf1/Cip1, p27Kip1, and p53 in IEC-6 cells. *Am J Physiol Cell Physiol* 276: pp. C684–91, 1999.

[330] Appella E, Anderson CW. Post-transcriptional modifications and activation of p53 by genotoxic stresses. *Eur J Biochem* 268: pp. 2764–72, 2001.

[331] Li L, Rao JN, Li J, Patel AR, Bass BL, Wang JY. Polyamine depletion stabilizes p53 resulting in inhibition of normal intestinal epithelial cell proliferation. *Am J Physiol Cell Physiol* 281: pp. C941–53, 2001.

[332] Zou T, Rao JN, Liu L, Marasa BS, Keledjian KM, Zhang AH, Xiao L, Wang JY. Polyamine depletion induces nucleophosmin modulating stability and transcriptional activity of p53 in intestinal epithelial cells. *Am J Physiol Cell Physiol* 289: pp. C686–96, 2005.

[333] ChanWY, Liu QR, Borjigin J, Busch H, Rennert OM, Tease LA, Chan PK. Characterization of the cDNA encoding human nucleophosmin and studies of its role in normal and abnormal growth. *Biochemistry* 28: pp. 1033–9, 1989.

[334] Ryder K, Lanahan A, Perez-Albuerne E, Nathans D. JunD: a third member of the jun gene family. *Proc Natl Acad Sci U S A* 86: pp. 1500–3, 1989.

[335] Hirai SI, Ryseck RP, Mechta F, Bravo K, Yaniv M. Characterization of junD: a new member of the jun proto-oncogene family. *EMBO J* 8: pp. 1433–9, 1989.

[336] Pfarr CM, Mechta F, Spyrou G, Lallemand D, Carillo S, Yaniv M. Mouse JunD negatively regulates fibroblast growth and antagonizes transformation by ras. *Cell* 76: pp. 747–60, 1994.

[337] Patel AR, Wang JY. Polyamine depletion is associated with an increase in JunD/AP-1 activity in small intestinal crypt cells. *Am J Physiol Gastrointest Liver Physiol* 276: pp. G441–50, 1999.

[338] Patel AR, Wang JY. Polyamines modulate transcription but not posttranscription of c-myc and c-jun in IEC-6 cells. *Am J Physiol Cell Physiol* 273: pp. C1020–9, 1997.

[339] Li L, Liu L, Rao JN, Esmaili A, Strauch ED, Bass BL, Wang JY. JunD stabilization results in inhibition of normal intestinal epithelial cell growth through p21 after polyamine depletion. *Gastroenterology* 123: pp. 764–79, 2002.

[340] Barnard JA, Warwick GJ, Gold LI. Localization of transforming growth factor-β isoforms in the normal murine small intestine and colon. *Gastroenterology* 105: pp. 67–73, 1993.

[341] Franzen P, Dijke PT, Ichijo H, Yamashita H, Schulz P, Heldin CH, Miyazono K. Cloning of a TGFβ type I receptor that forms a heteromeric complex with the TGFβ type II receptor. *Cell* 75: pp. 681–92, 1993.

[342] Attisano L, Wrana JL. Smads as transcriptional co-modulators. *Curr Opin Cell Biol* 12: pp. 235–43, 2000.

[343] Itoh S, Itoh F, Goumans MJ, Dijke PT. Signaling of transforming growth factor-β family members through Smad proteins. *Eur J Biochem* 267: pp. 6954–67, 2000.

[344] Xiao L, Rao JN, Zou T, Liu L, Marasa BS, Chen J, Turner DJ, Zhou H, Gorospe M, Wang JY. Polyamines regulate the stability of activating transcription factor-2 mRNA through RNA-binding protein HuR in intestinal epithelial cells. *Mol Biol Cell* 18: pp. 4579–90, 2007.

[345] Zhang X, Zou T, Rao JN, Liu L, Xiao L, Wang PY, Cui YH, Gorospe M, Wang JY. Stabilization of XIAP mRNA through the RNA binding protein HuR regulated by cellular polyamines. *Nucleic Acids Res* 37: pp. 7623–37, 2009.

[346] Wang PY, Rao JN, Zou T, Liu L, Xiao L, Yu TX, Turner DJ, Gorospe M, Wang JY. Post-transcriptional regulation of MEK-1 by polyamines through the RNA-binding protein HuR modulating intestinal epithelial apoptosis. *Biochem J* 426: pp. 293–306, 2010.

[347] Gorospe M. HuR in the mammalian genotoxic response. *Cell Cycle* 2: pp. 412–4, 2003.

[348] Brennan CM, Steitz JA. HuR and mRNA stability. *Cell Mol Life Sci* 58: pp. 266–77, 2001.

[349] Gherzi R, Lee KY, Briata P, Wegmuller D, Moroni C, Karin M Chen CY. KH domain RNA binding protein, KSRP, promotes ARE-directed mRNA turnover by recruiting the degradation machinery. *Mol Cell* 14: pp. 571–83, 2004.

[350] de Silanes IL, Zhan M, Lal A, Yang X, Gorospe M. Identification of a target RNA motif for RNA-binding protein HuR. *Proc Natl Acad Sci U S A* 101: pp. 2987–92, 2004.

[351] Heinonen M, Bono P, Narko K, Chang SH, Lundin J, Joensuu H, Furneaux H, Hla T, Haglund C, Ristimaki A. Cytoplasmic HuR expression is a prognostic factor in invasive ductal breast carcinoma. *Cancer Res* 65: pp. 2157–61, 2005.

[352] Zou T, Mazan-Mamczarz K, Rao JN, Liu L, Marasa BS, Zhang AH, Xiao L, Pullmann R, Gorospe M, Wang JY. Polyamine depletion increases cytoplasmic levels of RNA-binding protein HuR leading to stabilization of nucleophosmin and p53 mRNAs. *J Biol Chem* 281: pp. 19387–94, 2006.

[353] Zou T, Liu L, Rao JN, Marasa BS, Chen J, Xiao L, Zhou H, Gorospe M, Wang JY. Polyamines modulate the subcellular localization of RNA-binding protein HuR through AMP-activated protein kinase-regulated phosphorylation and acetylation of importin α1. *Biochem J* 409: pp. 389–98, 2008.

[354] Zou T, Rao JN, Liu L, Xiao L, Yu TX, Jiang P, Gorospe M, Wang JY. Polyamines regulate the stability of JunD mRNA by modulating the competitive binding of its 3' untranslated region to HuR and AUF1. *Mol Cell Biol* 30: pp. 5021–5032, 2010.

[355] Liu L, Rao JN, Zou T, Xiao L, Wang PY, Turner DJ, Gorospe M, Wang JY. Polyamines regulate c-Myc translation through Chk2-dependent HuR phosphorylation. *Mol Biol Cell* 20: pp. 4885–98, 2009.

Author Biographies

Rao N. Jaladanki received his Ph.D. in Endocrinology from Sri Venkateswara University, Tirupati, AP, India, in 1992. His research has been highly focused on gut physiology, particularly in the regulation of mucosal growth and repair under biological and various pathological conditions over the past fifteen years. The goal of Dr. Rao's research is to specifically define the roles and mechanisms of Ca^{2+} signaling and ion channels in gut epithelial cell proliferation, growth arrest, apoptosis, and migration. Dr. Rao is an Assistant Professor at the University of Maryland School of Medicine, and he has published more than 70 peer-reviewed papers and 7 book chapters and review articles.

Dr. Jian-Ying Wang is a Professor at the University of Maryland School of Medicine and a Senior Research Career Scientist at the US Department of Veterans Affairs. Dr. Wang is a leading investigator in the areas of gastrointestinal physiology and epithelial cell biology, with particular expertise in gut mucosal growth regulation and epithelial renewal. Dr. Wang's research has been highly focused on the roles and mechanisms of cellular polyamines in the regulation of gut epithelial cell proliferation, apoptosis, migration, and cell-to-cell interactions over the past two decades. He has published 122 peer-reviewed articles, 19 review articles and book chapters, and 3 books on the gut physiology and polyamine topics. Dr. Wang has been a consultant for various government agencies in both the US and China.